The Mycota

Edited by
K. Esser

The Mycota

I *Growth, Differentiation and Sexuality*
1st edition ed. by J.G.H. Wessels and F. Meinhardt
2nd edition ed. by U. Kües and R. Fischer
3rd edition ed. by J. Wendland

II *Genetics and Biotechnology*
1st and 2nd edition ed. by U. Kück

III *Biochemistry and Molecular Biology*
1st and 2nd edition ed. by R. Brambl and G. Marzluf
3rd edition ed. by D. Hoffmeister

IV *Environmental and Microbial Relationships*
1st edition ed. by D. Wicklow and B. Söderström
2nd edition ed. by C.P. Kubicek and I.S. Druzhinina
3rd edition ed. by I.S. Druzhinina and C.P. Kubicek

V *Plant Relationships*
1st edition ed. by G. Carroll and P. Tudzynski
2nd edition ed. by H.B. Deising

VI *Human and Animal Relationships*
1st edition ed. by D.H. Howard and J.D. Miller
2nd edition ed. by A.A. Brakhage and P.F. Zipfel

VII *Systematics and Evolution*
1st edition ed. by D.J. McLaughlin, E.G. McLaughlin, and P.A. Lemke†
2nd edition ed. by D.J. McLaughlin and J.W. Spatafora

VIII *Biology of the Fungal Cell*
1st and 2nd edition ed. by R.J. Howard and N.A.R. Gow
3rd edition ed. by D. Hoffmeister and M. Gressler

IX *Fungal Associations*
1st and 2nd edition ed. by B. Hock

X *Industrial Applications*
1st edition ed. by H.D. Osiewacz
2nd edition ed. by M. Hofrichter

XI *Agricultural Applications*
1st and 2nd edition ed. by F. Kempken

XII *Human Fungal Pathogens*
1st edition ed. by J.E. Domer and G.S. Kobayashi
2nd edition ed. by O. Kurzai

XIII *Fungal Genomics*
1st edition ed. by A.J.P. Brown
2nd edition ed. by M. Nowrousian

XIV *Evolution of Fungi and Fungal-Like Organisms*
Ed. by S. Pöggeler and J. Wöstemeyer

XV *Physiology and Genetics: Selected Basic and Applied Aspects*
1st edition ed. by T. Anke and D. Weber
2nd edition ed. by T. Anke and A. Schüffler

The Mycota

A Comprehensive Treatise on Fungi as Experimental Systems for Basic and Applied Research

Edited by K. Esser

VIII

Biology of the Fungal Cell
3rd Edition

Volume Editors:
Dirk Hoffmeister and Markus Gressler

Springer

Series Editor

Professor Dr. Dr. h.c. mult. Karl Esser
Allgemeine Botanik
Ruhr-Universität
44780 Bochum, Germany

Tel.: +49 (234)32-22211
Fax.: +49 (234)32-14211
e-mail: Karl.Esser@rub.de

Volume Editors

Professor Dr. Dirk Hoffmeister
Friedrich-Schiller-Universität
Pharmazeutische Mikrobiologie
Winzerlaer Strasse 2
07745 Jena, Germany

Tel.: +49 (3641)949851
Fax: +49 (3641)949852
e-mail: dirk.hoffmeister@leibniz-hki.de

Dr. Markus Gressler
Friedrich-Schiller-Universität
Pharmazeutische Mikrobiologie
Winzerlaer Strasse 2
07745 Jena, Germany

Tel.: +49 (3641) 949855
e-mail: Markus.Gressler@leibniz-hki.de

ISBN 978-3-030-05446-5 ISBN 978-3-030-05448-9 (eBook)
https://doi.org/10.1007/978-3-030-05448-9

Library of Congress Control Number: 2019933208

This Springer imprint is published by the registered company Springer Nature Switzerland AG
The registered company address is: Gewerbestrasse 11, 6330 Cham, Switzerland

Karl Esser

(born 1924) is retired Professor of General Botany and Director of the Botanical Garden at the Ruhr-Universität Bochum (Germany). His scientific work focused on basic research in classical and molecular genetics in relation to practical application. His studies were carried out mostly on fungi. Together with his collaborators he was the first to detect plasmids in higher fungi. This has led to the integration of fungal genetics in biotechnology. His scientific work was distinguished by many national and international honors, especially three honorary doctoral degrees.

Markus Gressler

(born 1984) gained a Ph.D. in biochemistry at the Friedrich Schiller University and the Leibniz Institute for Natural Product Research and Infection Biology—Hans Knöll Institute (HKI) in Jena (Germany) in 2015. From 2015 till 2017 he was a postdoctoral fellow at Institut Pasteur, Paris, before he pursued his research interests in 2017 at the Friedrich Schiller University, Jena, addressing the biosynthesis and the ecological aspects of polyketides and nonribosomal peptides in fungi.

Dirk Hoffmeister

(born 1972) is a mycologist and Full Professor for Pharmaceutical Microbiology at the Friedrich Schiller University Jena (Germany). He is also affiliated with the Leibniz Institute for Natural Product Research and Infection Biology—Hans Knöll Institute (HKI). Particularly emphasizing the basidiomycetes, research in his group is focused on the biochemical and genetic basis underlying fungal natural product biosyntheses.

Series Preface

Mycology, the study of fungi, originated as a sub discipline of botany and was a descriptive discipline, largely neglected as an experimental science until the early years of this century. A seminal paper by Blakeslee in 1904 provided evidence for self-incompatibility, termed "heterothallism," and stimulated interest in studies related to the control of sexual reproduction in fungi by mating-type specificities. Soon to follow was the demonstration that sexually reproducing fungi exhibit Mendelian inheritance and that it was possible to conduct formal genetic analysis with fungi. The names Burgeff, Kniep and Lindegren are all associated with this early period of fungal genetics research.

These studies and the discovery of penicillin by Fleming, who shared a Nobel Prize in 1945, provided further impetus for experimental research with fungi. Thus, began a period of interest in mutation induction and analysis of mutants for biochemical traits. Such fundamental research, conducted largely with Neurospora crassa, led to the one gene:one enzyme hypothesis and to a second Nobel Prize for fungal research awarded to Beadle and Tatum in 1958. Fundamental research in biochemical genetics was extended to other fungi, especially to *Saccharomyces cerevisiae*, and by the mid-1960s fungal systems were much favored for studies in eukaryotic molecular biology and were soon able to compete with bacterial systems in the molecular arena.

The experimental achievements in research on the genetics and molecular biology of fungi have benefited more generally studies in the related fields of fungal biochemistry, plant pathology, medical mycology, and systematics. Today, there is much interest in the genetic manipulation of fungi for applied research. This current interest in biotechnical genetics has been augmented by the development of DNA-mediated transformation systems in fungi and by an understanding of gene expression and regulation at the molecular level. Applied research initiatives involving fungi extend broadly to areas of interest not only to industry but to agricultural and environmental sciences as well.

It is this burgeoning interest in fungi as experimental systems for applied as well as basic research that has prompted publication of this series of books under the title The Mycota. This title knowingly relegates fungi into a separate realm, distinct from that of either plants, animals, or protozoa. For consistency throughout this series of volumes, the names adopted for major groups of fungi (representative genera in parentheses) are as follows:

Pseudomycota

Division:	Oomycota (*Achlya, Phytophthora, Pythium*)
Division:	Hyphochytriomycota

Eumycota

Division:	Chytridiomycota (*Allomyces*)
Division:	Zygomycota (*Mucor, Phycomyces, Blakeslea*)
Division:	Dikaryomycota
Subdivision:	Ascomycotina
Class:	Saccharomycetes (*Saccharomyces, Schizosaccharomyces*)
Class:	Ascomycetes (*Neurospora, Podospora, Aspergillus*)
Subdivision:	Basidiomycotina
Class:	Heterobasidiomycetes (*Ustilago, Tremella*)
Class:	Homobasidiomycetes (*Schizophyllum, Coprinus*)

We have made the decision to exclude from The Mycota the slime molds which, although they have traditional and strong ties to mycology, truly represent nonfungal forms insofar as they ingest nutrients by phagocytosis, lack a cell wall during the assimilative phase, and clearly show affinities with certain protozoan taxa.

The series throughout will address three basic questions: what are the fungi, what do they do, and what is their relevance to human affairs? Such a focused and comprehensive treatment of the fungi is long overdue in the opinion of the editors.

A volume devoted to systematics would ordinarily have been the first to appear in this series. However, the scope of such a volume, coupled with the need to give serious and sustained consideration to any reclassification of major fungal groups, has delayed early publication. We wish, however, to provide a preamble on the nature of fungi, to acquaint readers who are unfamiliar with fungi with certain characteristics that are representative of these organisms and which make them attractive subjects for experimentation.

The fungi represent a heterogeneous assemblage of eukaryotic microorganisms. Fungal metabolism is characteristically heterotrophic or assimilative for organic carbon and some nonelemental source of nitrogen. Fungal cells characteristically imbibe or absorb, rather than ingest, nutrients and they have rigid cell walls. The vast majority of fungi are haploid organisms reproducing either sexually or asexually through spores. The spore forms and details on their method of production have been used to delineate most fungal taxa. Although there is a multitude of spore forms, fungal spores are basically only of two types: (1) asexual spores are formed following mitosis (mitospores) and culminate vegetative growth, and (2) sexual spores are formed following meiosis (meiospores) and are borne in or upon specialized generative structures, the latter frequently clustered in a fruit body. The vegetative forms of fungi are either unicellular, yeasts are an example, or hyphal; the latter may be branched to form an extensive mycelium.

Regardless of these details, it is the accessibility of spores, especially the direct recovery of meiospores coupled with extended vegetative haploidy, that have made fungi especially attractive as objects for experimental research.

The ability of fungi, especially the saprobic fungi, to absorb and grow on rather simple and defined substrates and to convert these substances, not only into essential metabolites but into important secondary metabolites, is also noteworthy. The metabolic capacities of fungi have attracted much interest in natural products chemistry and in the production of antibiotics and other bioactive compounds. Fungi, especially yeasts, are important in fermentation processes. Other fungi are important in the production of enzymes, citric acid, and other organic compounds as well as in the fermentation of foods.

Fungi have invaded every conceivable ecological niche. Saprobic forms abound, especially in the decay of organic debris. Pathogenic forms exist with both plant and animal hosts. Fungi even grow on other fungi. They are found in aquatic as well as soil environments, and their spores may pollute the air. Some are edible; others are poisonous. Many are variously associated with plants as copartners in the formation of lichens and mycorrhizae, as symbiotic endophytes or as overt pathogens. Association with animal systems varies; examples include the predaceous fungi that trap nematodes, the microfungi that grow in the anaerobic environment of the rumen, the many insect associated fungi, and the medically important pathogens afflicting humans. Yes, fungi are ubiquitous and important. There are many fungi, conservative estimates are in the order of 100,000 species, and there are many ways to study them, from descriptive accounts of organisms found in nature to laboratory experimentation at the cellular and molecular level. All such studies expand our knowledge of fungi and of fungal processes and improve our ability to utilize and to control fungi for the benefit of humankind.

We have invited leading research specialists in the field of mycology to contribute to this series. We are especially indebted and grateful for the initiative and leadership shown by the Volume Editors in selecting topics and assembling the experts. We have all been a bit ambitious in producing these volumes on a timely basis and therein lies the possibility of mistakes and oversights in this first edition. We encourage the readership to draw our attention to any error, omission, or inconsistency in this series in order that improvements can be made in any subsequent edition.

Finally, we wish to acknowledge the willingness of Springer-Verlag to host this project, which is envisioned to require more than 5 years of effort and the publication of at least nine volumes.

Bochum, Germany
Auburn, AL, USA
April 1994

Karl Esser
Paul A. Lemke
Series Editors

Volume Preface

Since the turn of the twenty-first century, an exponential amount of fungal genomes has been sequenced and analyzed and are now available online with open access. However, it quickly turned out that the massive collection of in silico-generated data is not sufficient for biotechnological downstream application processes or for an in-depth understanding of interspecies cross-talks:

The current third edition of *The Mycota VIII* highlights aspects of fundamental fungal cell biology. The academic view on the field of fungal cell growth—long time denounced as a traditional but outranged field of fungal biology—has undergone a drastic change within the last 20 years. The arising need for fungal products necessitates the optimization of large-scale fermentation techniques which in turn requires a detailed knowledge in fungal development, cell division, cell wall production, and intracellular signal transduction: In food and pharmaceutical industries, the production of medically important fungal metabolites or food contaminants, such as the aflatoxins, is strictly dependent on the developmental stage of the producing fungi. In addition, some immune-modulating drugs are based on large fungal cell wall polysaccharides but due to their stickiness, their production is highly energy-wasting and complicated. However, phenotypic switches from filamentous to yeast-like growth forms ease not only the fermentation process but also the subsequent downstream processing of the fungal material. In sum, extended knowledge in fungal cell biology enables defined, precise cultivation conditions and therefore highly advances the white biotechnology in pharmaceutical and food industry.

From the ecological point of view, fungi are indispensable symbionts of most of higher plant roots supporting each other's growth. The tight fungus–plant interplay not only requires a shift of the fungal cellular shape, but also an adapted metabolism in response to defined extracellular signals. The discovery of fungal signal transduction pathways eases the understanding of the fine-tuned metabolic cross-flow between both symbionts and is therefore of increased interest in agriculture.

The editors are pleased to work with an enthusiastic group of international experts on the field of fungal cell biology who collected and combined the current knowledge, wrote the articles, and depicted main findings in illustrated figures. The close collaboration with the editors enabled a holistic view on fungal cell division and development covering the two most important divisions of fungi, ascomycetes and basidiomycetes. The volume highlights further aspects of cell–cell connections, cell shape switches, polarized growth, and protein transport within the cells. As a special feature, the current volume spans the divide between fungal cell growth and signal transmission/transduction and shows its interconnectivity.

The editors thank Steven D. Harris, Ph.D., who initiated this volume. We also cordially thank Dr. Andrea Schlitzberger of Springer Nature for the discussion and coordination of this current volume of *The Mycota*. The editors are grateful to benefit from the profound experience of Senior Editor, Professor Emeritus Karl Esser, and thank him for his kind and encouraging advice.

Jena, Germany
November 2018

Markus Gressler
Dirk Hoffmeister

Contents

List of Contributors

REINHARD FISCHER
Department of Microbiology, Institute for Applied Biosciences, Karlsruhe Institute of Technology (KIT), Karlsruhe, Germany

KAP-HOON HAN
Department of Pharmaceutical Engineering, Woosuk University, Wanju, Republic of Korea

YVONNE HEILIG
Institute for Biology II – Molecular Plant Physiology, Albert-Ludwigs University Freiburg, Freiburg, Germany

HECHUN JIANG
Jiangsu Key Laboratory for Microbes and Functional Genomics, Jiangsu Engineering and Technology Research Center for Microbiology, College of Life Sciences, Nanjing Normal University, Nanjing, China

MIN-JU KIM
School of Food Science and Biotechnology, Institute of Agricultural Science and Technology, Kyungpook National University, Daegu, Republic of Korea

KATSUHIKO KITAMOTO
Department of Biotechnology, The University of Tokyo, Tokyo, Japan

MI-KYUNG LEE
Biological Resource Center (BRC), Korea Research Institute of Bioscience and Biotechnology (KRIBB), Jeongeup-si, Republic of Korea

ROGER R. LEW
Department of Biology, York University, Toronto, Canada

LING LU
Jiangsu Key Laboratory for Microbes and Functional Genomics, Jiangsu Engineering and Technology Research Center for Microbiology, College of Life Sciences, Nanjing Normal University, Nanjing, China

Francis Martin
Laboratoire d'excellence ARBRE, Centre INRA de Nancy-Lorraine, UMR INRA/ Université de Lorraine 1136 'Interactions Arbres/Microorganismes', Champenoux, France

Jun-ichi Maruyama
Department of Biotechnology, The University of Tokyo, Tokyo, Japan

Hee-Soo Park
School of Food Science and Biotechnology, Institute of Agricultural Science and Technology, Kyungpook National University, Daegu, Republic of Korea

Clément Pellegrin
Laboratoire d'excellence ARBRE, Centre INRA de Nancy-Lorraine, UMR INRA/ Université de Lorraine 1136 'Interactions Arbres/Microorganismes', Champenoux, France

Stephan Seiler
Institute for Biology II – Molecular Plant Physiology, Albert-Ludwigs University Freiburg, Freiburg, Germany

Freiburg Institute for Advanced Studies (FRIAS), Albert-Ludwigs University Freiburg, Freiburg, Germany

Norio Takeshita
Faculty of Life and Environmental Sciences, University of Tsukuba, Tsukuba, Japan

Claire Veneault-Fourrey
Laboratoire d'excellence ARBRE, Centre INRA de Nancy-Lorraine, UMR INRA/ Université de Lorraine 1136 'Interactions Arbres/Microorganismes', Champenoux, France

Jae-Hyuk Yu
Department of Bacteriology, University of Wisconsin, Madison, WI, USA

Department of Systems Biotechnology, Konkuk University, Seoul, Republic of Korea

Yuanwei Zhang
Jiangsu Key Laboratory for Microbes and Functional Genomics, Jiangsu Engineering and Technology Research Center for Microbiology, College of Life Sciences, Nanjing Normal University, Nanjing, China

Fungal Cell Growth

The Woronin Body: A Fungal Organelle Regulating Multicellularity

Jun-ichi Maruyama[1], Katsuhiko Kitamoto[1]

CONTENTS

I. Multicellularity of Filamentous Fungi

Filamentous fungi form straight hyphae via polarized tubular extension of the hyphal tip and consist of a network of straight primary hyphae with the formation of branches. Hyphae are compartmentalized into distinct cells by the formation of a septum, which is not frequently observed in early-diverging fungi, such as members of the phyla *Mucoromycota* and *Chytridiomycota*, but is regularly found in later-diverging fungi belonging to the phyla *Ascomycota* and *Basidiomycota* (Jedd 2011). The septum is proposed to increase the mechanical integrity of hyphae and to divide the mycelium into sections of distinct growth states, such as mitotic and non-mitotic cells (Fiddy and Trinci 1976; Momany et al. 2002; Nayak et al. 2010; Edgerton-Morgan and Oakley 2012). However, the septum does not completely separate hyphae due to the presence of a septal pore, which is a perforated structure that allows the exchange of the cytoplasmic constituents, including organelles, between adjacent hyphal cells (Lew 2005; Tey et al. 2005; Ng et al. 2009; Bleichrodt et al. 2015). Such cell-to-cell connectivity resembles gap junctions in animal cells and plasmodesmata in plant cells.

Cell-to-cell connectivity through the septal pore is associated with the catastrophic risk of cytoplasmic loss by cells adjacent to individually damaged hyphae. This risk was demonstrated by the physical damage of hyphae with a razor blade (Trinci and Collinge 1974) or pulse laser (Lichius et al. 2012), treatment with cell wall-destabilizing reagents (Bowman et al. 2002), and exposure to hypotonic shock (Fig. 1a) (Jedd and Chua 2000; Maruyama et al. 2005). Cell death is induced by aging and heterokaryon incompatibility (Fleißner and Glass 2007). Despite such damage, the cytoplasm of cells immediately adjacent to the wounded cell is typically retained (Fig. 1b) (Jedd and Chua 2000; Maruyama et al. 2005). The protected cells then initiate regrowth of the mycelium by producing a new hyphal tip (Jedd and Chua 2000; Maruyama et al. 2006; Maruyama and Kitamoto 2007; Lichius et al. 2012). These processes represent an inherent defense system for promoting survival by preventing the simultaneous loss of cytoplasm from multiple cells upon hyphal wounding. Hence, the septal pore is an important subcellular structure for maintaining the multicellularity of filamentous fungi.

[1] Department of Biotechnology, The University of Tokyo, Tokyo, Japan; e-mail: amarujun@mail.ecc.u-tokyo.ac.jp

Biology of the Fungal Cell, 3rd Edition
The Mycota VIII
D. Hoffmeister, M. Gressler (Eds.)

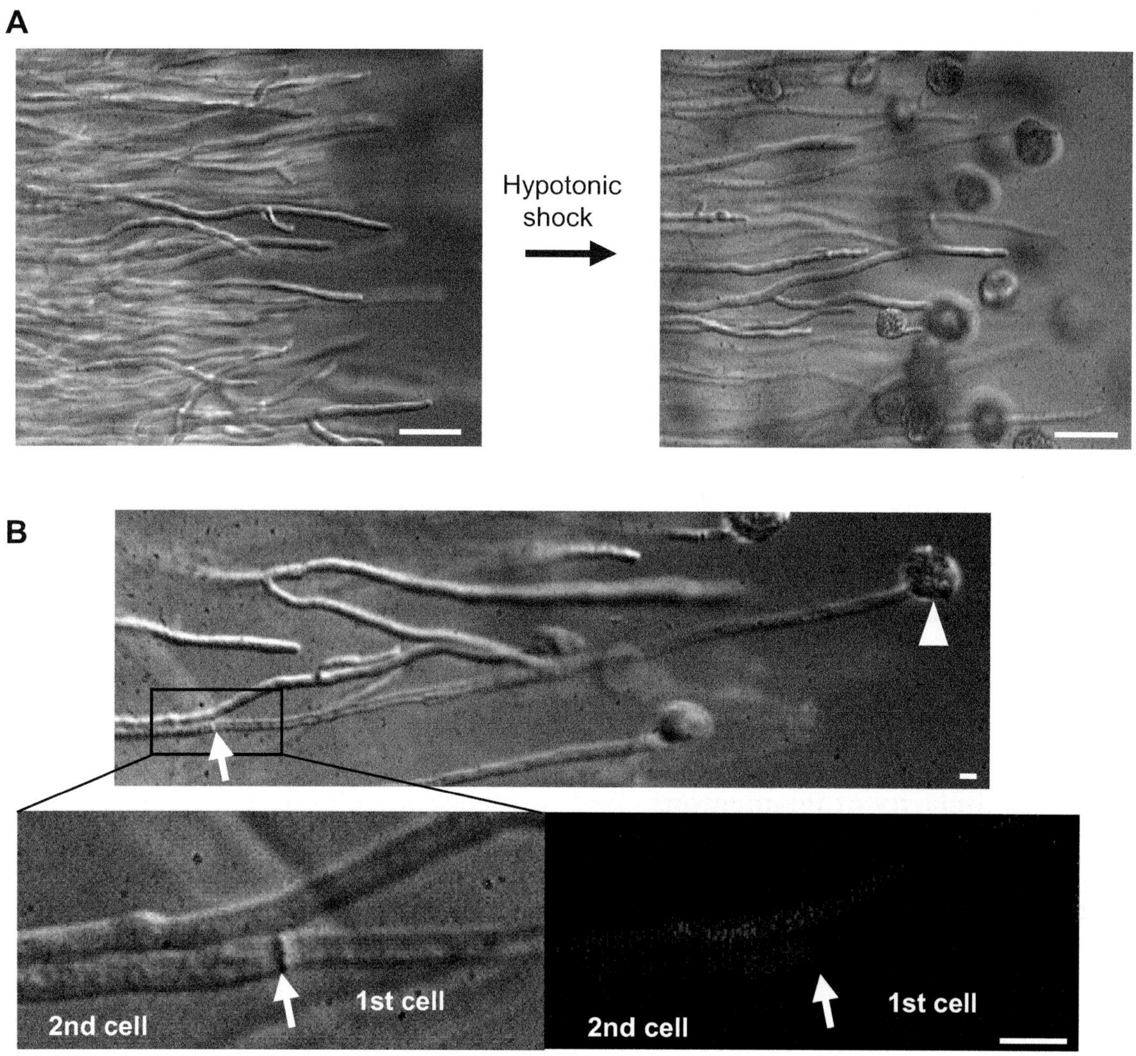

Fig. 1 Hyphal wounding and the prevention of excessive cytoplasmic loss. (**a**) Hyphal tip bursting upon hypotonic shock. Hyphal tips at the margin of an *A. oryzae* colony grown on agar medium were observed by differential interference contrast (DIC) microscopy before and after flooding hyphae with water. Bar: 50 μm. (**b**) Prevention of the excessive loss of cytoplasm upon hyphal wounding. The cytoplasm is labeled by expressing EGFP. An arrowhead and arrow indicate a burst hyphal tip and the adjacent septum, respectively. Note that the cell (2nd) adjacent to the lysed cell (1st) retains its cytoplasmic constituents, as determined by DIC and fluorescence microcopy. Bar: 10 μm

II. The Woronin Body, an Organelle Specific to *Pezizomycotina*

Filamentous fungal species that form septa have evolved to possess specialized membrane-bound organelles located around the septum; the *Pezizomycotina* (*Ascomycota*) contain structures called the **Woronin body** (Markham and Collinge 1987), and *Agaricomycotina* (*Basidiomycota*) have an endoplasmic reticulum (ER)-related septal pore cap (Müller et al. 1998). Woronin bodies were first identified as highly refractive particles located near the septum in the filamentous ascomycete *Ascobolus pulcherrimus* by the Russian mycologist Michael Stepanovitch Woronin (Woronin 1864). The organelle was later named by Buller

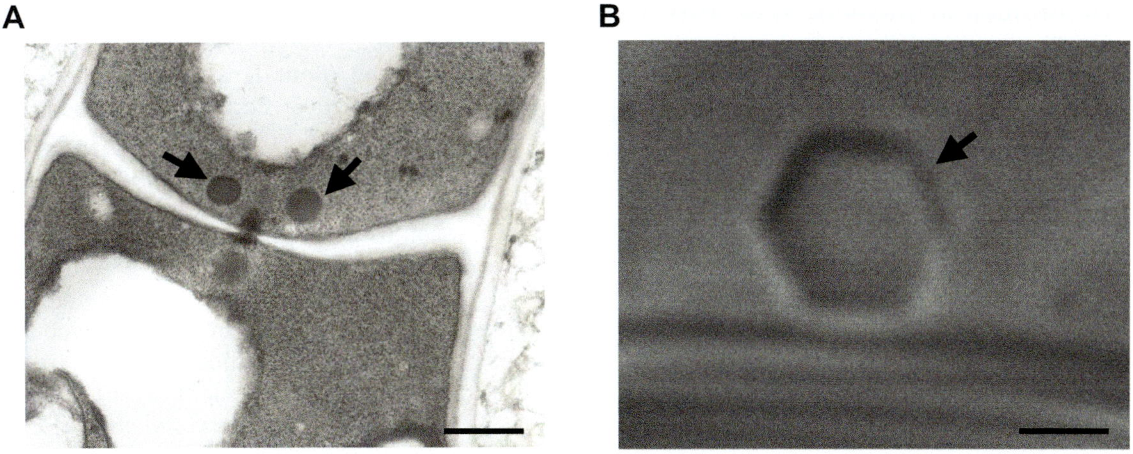

Fig. 2 Morphology of the Woronin body. (a) Transmission electron microscopic observation of Woronin bodies (arrows) near the septum in *A. oryzae*. Bar: 500 nm. (b) Differential interference contrast (DIC) image of a hexagonal Woronin body localized to the cell cortex in *N. crassa*. Woronin bodies are indicated by arrows. Bar: 1 μm

(1933) in recognition of Woronin's discovery and is restricted to *Pezizomycotina* species (Markham and Collinge 1987). The Woronin body has two morphologically distinct subclasses; it predominantly appears by transmission electron microscopy as a spherical electron-dense structure near the septum (Fig. 2a), although in a small number of species, such as *Neurospora crassa*, Woronin bodies form hexagonal crystalline structures that are occasionally visible by light microscopy (Fig. 2b) (Markham 1994). By both electron and light microscopy, Woronin bodies were observed to plug the septal pore upon hyphal wounding (Markham and Collinge 1987). Therefore, it was long suspected that the primary function of Woronin bodies was the prevention of excessive cytoplasmic loss from cells adjacent to damaged or lysed cells (Fig. 3a).

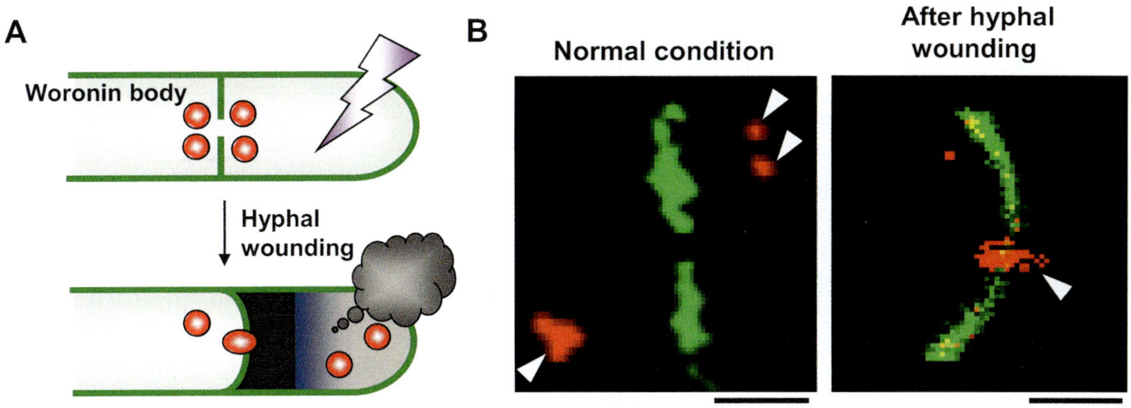

Fig. 3 Woronin body function. (a) Schematic representation of the septal plugging function of Woronin bodies. (b) Confocal images of Woronin bodies (red, arrowheads) and septa (green) before (left) and after (right) hyphal wounding. Woronin bodies and septa were labeled by DsRed2 and EGFP, respectively (Maruyama et al. 2005). Bar: 2 μm

III. Molecular Basis of Woronin Body Structure

Although Woronin bodies were discovered approximately 150 years ago, the composition of Woronin bodies long remained unclear. Protease digestion of ultrathin sections suggested that Woronin bodies contained proteinaceous materials (McKeen 1971; Mason and Crosse 1975). Woronin bodies from *N. crassa* were first purified in 2000 through differential and density gradient centrifugation, allowing for the identification of **Hex1** as a **major structural protein** (Jedd and Chua 2000; Tenney et al. 2000). Genes encoding Hex1 are conserved in *Pezizomycotina* species (Jedd and Chua 2000; Asiegbu et al. 2004; Curach et al. 2004; Soundararajan et al. 2004; Maruyama et al. 2005; Beck and Ebel 2013). Deletion of the *hex1* gene results in defective Woronin body formation and severe cytoplasmic bleeding upon hyphal wounding (Jedd and Chua 2000; Maruyama et al. 2005; Beck and Ebel 2013). When the apical neighboring cell is selectively ruptured, the subapical cells reinitiate hyphal growth by producing a new branch, but deletion of the *hex1* gene severely reduces or completely loses the activity of growth reinitiation (Tegelaar and Wösten 2017).

Characterization of Hex1 protein revealed that it spontaneously self-assembles to form a solid core, thereby providing mechanical resistance against the protoplasmic streaming pressure that is generated upon hyphal wounding (Jedd and Chua 2000; Yuan et al. 2003). The crystal structure of Hex1 consists of three intermolecular contacts that promote self-assembly (Yuan et al. 2003). Interestingly, the structural properties of Hex1 resemble those of eukaryotic translation initiation factor 5A (eIF5A) proteins (Kim et al. 1998; Peat et al. 1998), suggesting that the *hex1* gene evolved from *eIF-5A* by gene duplication in the ancestor of *Pezizomycotina*. Phosphorylation of Hex1 also contributes to the formation of the multimeric core of the Woronin body (Tenney et al. 2000; Juvvadi et al. 2007).

The *N. crassa hex1* gene encodes a single translational product (Jedd and Chua 2000), although alternative splicing of the *hex1* gene was reported in several species, including *Magnaporthe grisea*, *Aspergillus oryzae*, and *Aspergillus fumigatus* (Soundararajan et al. 2004; Maruyama et al. 2005; Beck et al. 2013). Both spliced and non-spliced transcripts yield two polypeptides that sediment in high-density protein fractions (Maruyama et al. 2005), suggesting that these peptides participate in the self-assembly and formation of the Woronin body core matrix. Moreover, it was demonstrated that a poly-histidine motif encoded within an alternatively spliced region targets Hex1 to the septal pore (Beck et al. 2013).

Although the hexagonal crystal structure of the *N. crassa* Woronin body is easily observed by light microscopy, the fusion of Hex1 with a fluorescent protein enables the septal plugging activity of Woronin bodies to be visualized in most *Pezizomycotina* species (Fig. 3b) (Maruyama et al. 2005; Beck and Ebel 2013). Using this approach, Bleichrodt et al. (2012) reported that the Woronin body reversibly plugs the septal pore during normal growth, a characteristic that contrasts the conventional view of this organelle functioning in wound healing. The analysis of gene expression activity for individual hyphae revealed that although wild-type cells exhibited heterogeneous activity, cells lacking Woronin bodies through deletion of *hex1* (Δ*hex1* strain) displayed more uniform gene expression activity (Bleichrodt et al. 2012). Thus, it was proposed that Woronin bodies contribute to the generation of hyphal populations with different cellular activities by occasionally preventing cell-to-cell connectivity via the plugging of the septal pore. The generation of such colonial heterogeneity under normal growth conditions may protect cells against environmental stresses, as evidenced by the sensitivity of the *A. oryzae* Δ*hex1* strain to heat stress (Bleichrodt et al. 2012). Thus, it appears that Woronin bodies have a gatekeeper role in regulating cell-to-cell channels by simple plugging of septal pore, similar to that observed upon hyphal wounding (Jedd and Pieuchot 2012).

In addition to septal plugging, Hex1 has other physiological impacts at the cellular level. For example, deletion of the *hex1* gene results in defective conidiation (asexual spore formation) (Yuan et al. 2003; Son et al. 2013), impaired growth under nitrogen starvation (Soundararajan et al. 2004), and increased sensitivity to cell wall and membrane-destabilizing agents (Beck et al. 2013). In plant pathogenic fungi, Hex1 is also required for efficient pathogenesis and viral RNA accumulation (Soundararajan et al. 2004; Son et al. 2013). These lines of evidence suggest additional physiological functions of Woronin bodies, which need further investigation of the action mechanisms.

IV. Woronin Body Biogenesis from Peroxisomes

Electron microscopic studies examining the subcellular origin of Woronin bodies demonstrated the inclusion of these organelles within **microbodies** (Wergin 1973; Camp 1977). This finding was supported by the specific binding of antibodies against microbody-specific signal peptides to Woronin bodies (Keller et al. 1991). A relationship between peroxisomes and the Woronin body was clearly indicated by the finding that the C-terminus of Hex1 protein contains peroxisomal targeting signal sequence 1 (PTS1) (Jedd and Chua 2000). In addition, time-lapse imaging analysis revealed that Woronin bodies bud from peroxisomes (Tey et al. 2005).

Woronin body formation occurs at the hyphal apex through a process involving apically biased expression of the *hex1* gene (Tey et al. 2005). Woronin body biogenesis requires peroxins that mediate the import of peroxisomal matrix and membrane proteins (Fig. 4a) (Managadze et al. 2007; Liu et al. 2008; Li et al. 2014). Specifically, Hex1 associates with Pex26, a peroxin that is enriched in Woronin body-containing peroxisomes and recruits the AAA ATPases Pex1 and Pex6 to the peroxisomal membrane for receptor recycling (Liu et al. 2011). The recruitment of Pex26 leads to the formation of a parallel activation loop for Pex5 recycling and Hex1 import, allowing for the efficient biogenesis of Woronin bodies. Fam1 was originally identified in *Colletotrichum orbiculare* as a *Pezizomycotina*-specific ortholog of Pex22, which functions in the recycling of PTS receptors from peroxisomes to the cytosol (Fig. 4a). Fam1 is specifically localized on the membrane of Woronin bodies, raising the possibility that recycling of PTS receptor occurs in Woronin bodies (Kubo et al. 2015).

In addition, the budding of Woronin bodies from peroxisomes appears to require dynamin-related proteins (Würtz et al. 2008). The peroxisome proliferator protein Pex11 is needed for the differentiation of Woronin bodies from peroxisomes (Fig. 4b) and also affects the septal plugging function of these organelles (Escaño et al. 2009). The genetic screening of *N. crassa* mutants defective in Woronin body biogenesis identified several peroxins and the Woronin body sorting complex (WSC) protein that were critical for this process (Liu et al. 2008). WSC is a *Pezizomycotina*-specific protein that recruits the Hex1 assembly to the matrix side of the peroxisomal membrane and facilitates budding of the Woronin body (Fig. 4b) (Liu et al. 2008). WSC belongs to the PMP22 (peroxisome membrane protein)/MPV17 (myeloproliferative leukemia virus 17) gene family and is proposed to have evolved to possess new functional properties, including self-assembly and Hex1 binding, from ancestral PMP22 (Jedd 2011).

ApsB, a component of the microtubule-organizing center (MTOC) that has functional peroxisomal targeting signal sequence 2 (PTS2), interacts with Hex1 (Zekert et al. 2010). In addition, TmpL, a transmembrane protein involved in redox-related signal transduction, is localized to Woronin bodies (Kim et al. 2009). However, it is unclear how ApsB and TmpL are involved in Woronin body biogenesis and function.

V. Septal Tethering of the Woronin body

In most *Pezizomycotina* species, Woronin bodies are typically tethered to the septum by a filament at a distance of 100–200 nm (Momany et al. 2002; Maruyama et al. 2005). The tethering structure was shown to have elastic properties, as demonstrated by the rapid return of the Woronin bodies to the septum after physical separation by laser trapping (Berns et al.

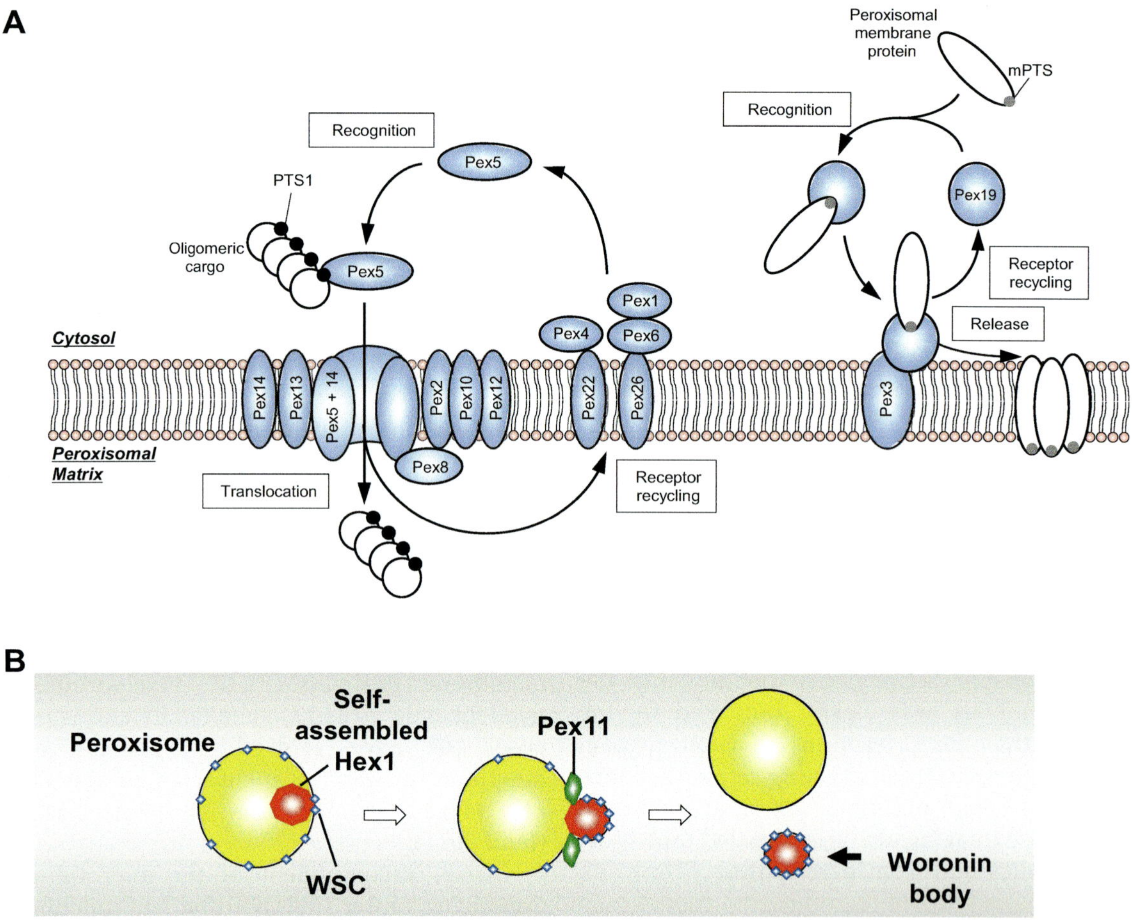

Fig. 4 Woronin body biogenesis from peroxisomes. (**a**) Schematic diagrams of the import machineries of peroxisomal matrix and membrane proteins. Peroxins labeled in blue were reported to be involved in Woronin body function. *PTS1* peroxisomal targeting signal 1, *mPTS* membrane peroxisome targeting signal. (**b**) Model of Woronin body biogenesis from peroxisomes

1992). Moreover, Woronin bodies rapidly insert into the septal pore upon hyphal wounding and also reversibly plug the septal pore during normal growth (Bleichrodt et al. 2012). These processes require the proper positioning of the Woronin body and sufficient flexibility of the tethering linker.

In contrast, the Woronin bodies of a few members of the genera *Neurospora* and *Sordaria* are associated with the cortex in a delocalized pattern (Plamann 2009). In *N. crassa*, newly synthesized Hex1 proteins are imported into peroxisomes in the apical compartment (Tey et al. 2005), and the newly formed Woronin bodies are then inherited into subapical compartments via association with the cell cortex (Tey et al. 2005; Liu et al. 2008). During a screening for mutants that accumulate Woronin bodies in the apical compartment, Ng et al. (2009) identified the *leashin* locus, which is comprised of the adjacent genes *lah-1* and *lah-2*. The N-terminal region of LAH-1 binds to Woronin bodies via the membrane protein WSC, whereas the C-terminal region mediates the association of Woronin bodies with the cell cortex (Fig. 5a) (Ng et al. 2009). In contrast, LAH-2, which localizes to the hyphal tip and septum, is not involved in Woronin body function (Fig. 5a). In *N. crassa* cells expressing an LAH-1/LAH2 fusion construct, Woronin bodies

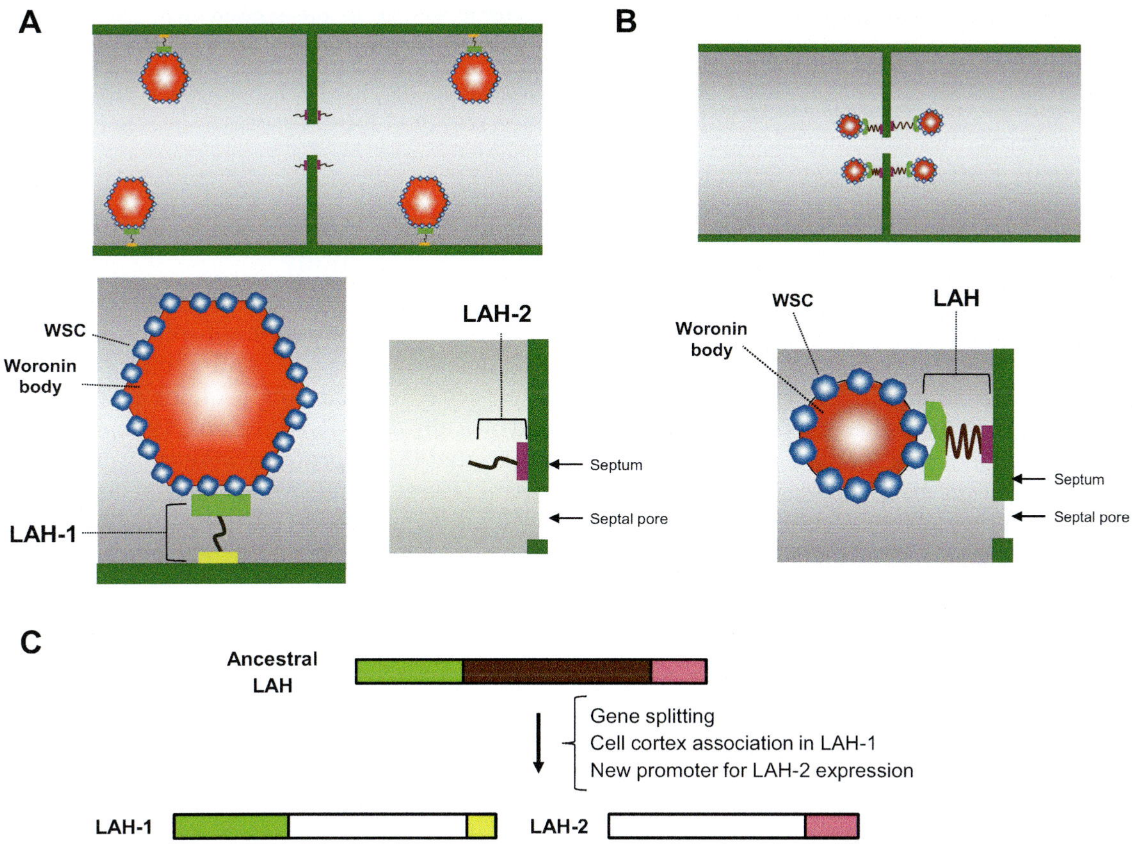

Fig. 5 Subcellular localization of Woronin bodies and tethering proteins. (**a**) Delocalized pattern of cortex-associated Woronin bodies via LAH-1 protein. (**b**) Septal tethering of Woronin bodies via LAH protein. (**c**) Illustration of the generation of two LAH proteins from the ancestral LAH protein

accumulate at both sides of the septum, which only has partial ability to prevent the excessive loss of cytoplasm upon hyphal injury (Ng et al. 2009). This finding indicates an additional requirement of the tethered Woronin bodies for **septal pore plugging**.

Other species, such as *A. fumigatus* and *A. oryzae*, with tethered Woronin bodies have large **LAH proteins** consisting of a single polypeptide of over 5000 amino acids (Beck et al. 2013; Han et al. 2014). LAH is required for the tethering of Woronin bodies to the septum and is involved in the Woronin body function of preventing the excessive loss of cytoplasm but to a lesser extent than the major Woronin body protein Hex1 (Han et al. 2014). This property may be explained by the fact that untethered Woronin bodies are able to plug the septal pore, but not as quickly as tethered ones.

Large LAH protein can be functionally divided into conserved N- and C-terminal regions and a non-conserved central region (Beck et al. 2013; Han et al. 2014): The N-terminal region associates with Woronin bodies in a WSC-dependent manner, and the C-terminal region containing a transmembrane spanning region mediates localization of LAH to the septum (Fig. 5b) (Beck et al. 2013; Han et al. 2014; Leonhardt et al. 2017). A truncated LAH protein consisting of only the N- and C-terminal regions retains the ability to tether Woronin bodies to the septum; however, the Woronin bodies are located closer (~50 nm) to the septum than those in wild-type cells (Han et al. 2014). This difference is roughly consistent with the 70-nm length of the approximately 2700-amino-acid central region of LAH, as was estimated based on the length

(1 μm) of the 4-mDa protein titin (Nave et al. 1989). The elasticity of the Woronin body tether that was previously demonstrated by laser-capture experiments (Berns et al. 1992) is conferred by the central non-conserved region of LAH (Han et al. 2014). The non-conserved central region in LAH is predicted to be disordered (Han et al. 2014) and presumably functions as a molecular spring, similarly to the muscle protein titin, which exhibits molecular spring-like elasticity via its intrinsically disordered region (Li et al. 2001). Moreover, in cells expressing LAH protein lacking the central region, tethered Woronin bodies do not plug the septal pore, even after hyphal wounding (Han et al. 2014). Collectively, efficient septal plugging requires not only that Woronin bodies are tethered to the septum but also that the tether must have sufficient elasticity to allow for the relatively unrestricted movement of Woronin bodies. The occasional observation that Woronin bodies from the ruptured cell side plug the septal pore suggests an active mechanism of Woronin body movement (Steinberg et al. 2017a). Surprisingly, ATP depletion by the respiration inhibitor carbonyl cyanide *m*-chlorophenylhydrazine induces the translocation of Woronin bodies into the septal pore without cell wounding. This suggests that ATP is required to prevent the septal plugging by Woronin bodies, leading to the speculation that ATP may bind to the tethering protein LAH for preventing a conformational change and contraction of the protein (Steinberg et al. 2017b).

The molecular origin of LAH remains elusive, but this protein first appeared in the clade of *Pezizomycotina* species together with Woronin bodies (Jedd 2011). The ancestral *lah* gene encodes a single, large polypeptide and was subsequently split into two genes, *lah1* and *lah2*, in the *Neurospora-Sordaria* clade. The *lah1* gene acquired an early termination sequence and a new cortex association domain at the C-terminus and a new promoter controlled expression of the downstream *lah2* gene (Fig. 5c). Thus, as a result of the gene splitting of the *lah* gene, the localization of the Woronin body was shifted from septal tethering to the cortex. *N. crassa* and *Sordaria fimicola* exhibit extensive cytoplasmic streaming through septal pores (Lew 2005; Tey et al. 2005; Ng et al. 2009) and have unusually rapid growth rates (>1 μm/s) (Ryan et al. 1943). As the tethering of Woronin bodies to the septum may have reduced the exchange of cytoplasm between adjacent cells, a delocalized pattern of Woronin bodies with cortex association may have been selected to support the rapid growth of these species.

VI. Proteins Associated with the Woronin Body

Through the purification of Woronin body-associated proteins and bioinformatics approaches for detecting proteins that contain intrinsically disordered regions, 17 **septal pore-associated (SPA) proteins** were identified in *N. crassa* (Lai et al. 2012). Intrinsically disordered proteins range from partially to completely unstructured and lacking globular folds; however, these proteins are capable of folding upon binding to target molecules (Wright and Dyson 2009). The number of reports related to intrinsically disordered proteins has increased significantly in recent years, and many biological functions of these proteins have been revealed (Oldfield and Dunker 2014). The loss-of-function of several SPA proteins leads to excessive septation, septal pore degeneration, and uncontrolled Woronin body activation (Lai et al. 2012). Spa10 protein is required for the localization of LAH C-terminal region at the mature septal pore, consequently stable tethering of Woronin bodies to the septum (Leonhardt et al. 2017). These findings suggest that the septal pore is a complex subcellular site for the assembly of unstructured proteins, which contribute to diverse states of cell-to-cell connectivity.

The *Pezizomycotina*-specific protein SO, which was originally found in *N. crassa* and was named based on the "soft" appearance of the mutant colony, is important for hyphal fusion and sexual reproduction (Fleißner et al. 2005; Engh et al. 2007). SO protein consists of approximately 1200 amino acids and contains a single WW domain. Pro40, a *Sordaria macrospora* SO homolog, functions as a scaffold by associating with protein kinases involved in cell wall integrity (Teichert et al. 2014). SO protein is dispersed throughout the cytoplasm under normal growth conditions but accumulates at

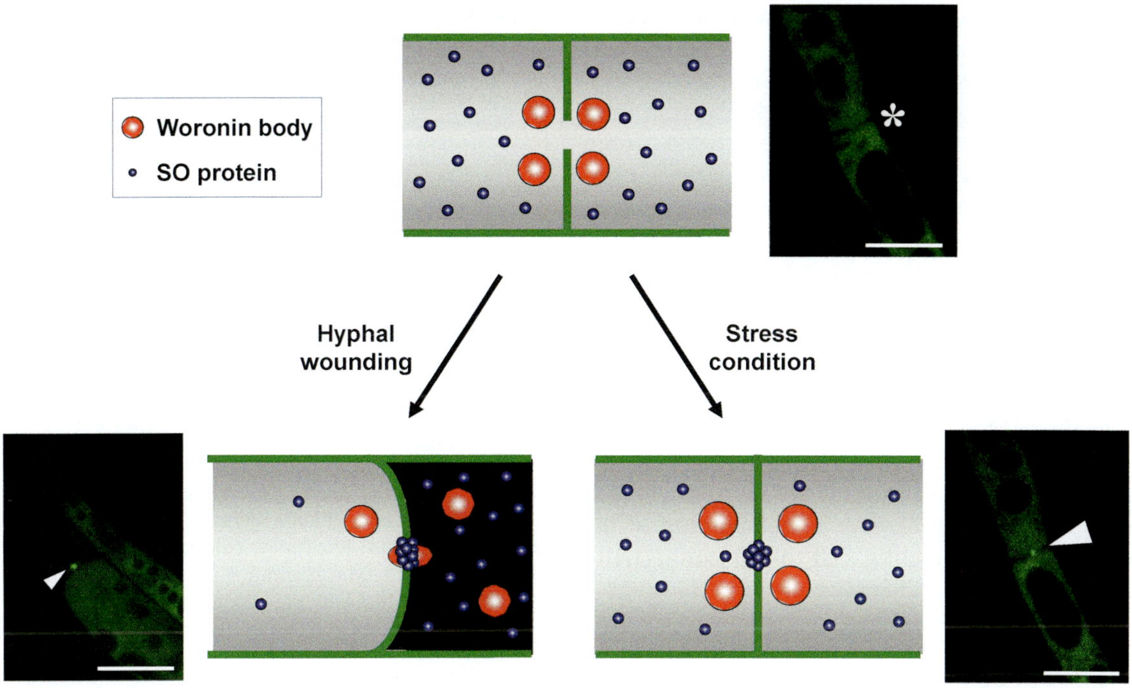

Fig. 6 Accumulation of SO protein at the septal pore. SO localization was visualized by expressing as EGFP fusion protein. (Left) Upon hyphal wounding, SO accumulates at the septal pore together with the Woronin body. (Right) SO also accumulates at the septal pore under stressed conditions. Asterisk indicates the septum, and the arrowhead indicates the accumulation of SO protein at the septal pore. Bars: 5 μm

the septal pore adjacent to wounded cells (Fig. 6), as was reported in *N. crassa* and *A. oryzae* (Fleißner and Glass 2007; Maruyama et al. 2010). *S. macrospora* Pro40 protein colocalizes with Woronin bodies (Engh et al. 2007). Deletion of the *so* gene delays septal plugging and reduces the number of hyphae that prevent excessive cytoplasmic loss in *N. crassa* and *A. oryzae* (Fleißner and Glass 2007; Maruyama et al. 2010). Taken together, these findings indicate that SO protein has septal plugging activity by accumulating at the septal pore, similar to the function of Woronin bodies. Additionally, SO protein accumulates at the septal pore in aging hyphae (Fleißner and Glass 2007) and under various stress conditions, including low/high temperature, high acidity/alkalinity, and nitrogen/carbon depletion (Fig. 6) (Maruyama et al. 2010). In response to pulse laser treatment, which physically stresses cells without causing hyphal wounding, SO rapidly accumulates at the septal pore nearest to the stressed hyphal area (Maruyama et al. 2010). Thus, SO protein may regulate cell-to-cell connectivity via the septal pore in a stress-dependent manner.

A study of *Aspergillus nidulans* revealed that mitosis interrupts cell-to-cell connectivity through the septal pore (Shen et al. 2014). The mitotic NIMA kinase was found to be localized to the septum during interphase and to contribute to keeping the septal pore open but is translocated into the nucleus to initiate mitosis, resulting in septal closure. Notably, however, the mitotic regulation of cell-to-cell connectivity is independent of Woronin bodies and SO protein. Identification of the NIMA kinase substrate(s) may provide insight into the molecular mechanisms underlying mitotic interruption of cell-to-cell connectivity.

VII. Conclusions and Perspectives

Despite the discovery of Woronin bodies over 150 years ago, our understanding of these unique fungal organelles was exclusively based on primitive microscopic studies conducted in the twentieth century. However, in the past 20 years, molecular structure and function of Woronin bodies have been largely elucidated. Genomic changes such as gene duplication and splitting generated and modified the Woronin body to allow for specific multicellular organization of individual *Pezizomycotina* species. Although the primary function of Woronin bodies had long been thought to be wound healing, recent molecular studies have revealed new physiological functions of these organelles. To date, however, the molecular machineries controlling and mediating septal closure and cell repair remain mostly unknown. Further investigations to functionally link Woronin bodies and related molecules in the vicinity of the septum will lead to a more comprehensive understanding of the mechanisms regulating cell-to-cell connectivity and fungal multicellularity.

References

Asiegbu FO, Choi W, Jeong JS, Dean RA (2004) Cloning, sequencing and functional analysis of *Magnaporthe grisea MVP1* gene, a *hex-1* homolog encoding a putative "Woronin body" protein. FEMS Microbiol Lett 230:85–90

Beck J, Ebel F (2013) Characterization of the major Woronin body protein HexA of the human pathogenic mold *Aspergillus fumigatus*. Int J Med Microbiol 303:90–97

Beck J, Echtenacher B, Ebel F (2013) Woronin bodies, their impact on stress resistance and virulence of the pathogenic mould *Aspergillus fumigatus* and their anchoring at the septal pore of filamentous Ascomycota. Mol Microbiol 89:857–871

Berns MW, Aist JR, Wright WH, Liang H (1992) Optical trapping in animal and fungal cells using a tunable, near-infrared titanium-sapphire laser. Exp Cell Res 198:375–378

Bleichrodt RJ, van Veluw GJ, Recter B, Maruyama J, Kitamoto K, Wösten HA (2012) Hyphal heterogeneity in *Aspergillus oryzae* is the result of dynamic closure of septa by Woronin bodies. Mol Microbiol 86:1334–1344

Bleichrodt RJ, Hulsman M, Wösten HA, Reinders MJ (2015) Switching from a unicellular to multicellular organization in an *Aspergillus niger* hypha. MBio 6:e00111

Bowman JC, Scott Hicks P, Kurtz MB, Rosen H, Schmatz DM, Liberator PA, Douglas CM (2002) The antifungal echinocandin caspofungin acetate kills growing cells of *Aspergillus fumigatus in vitro*. Antimicrob Agents Chemother 46:3001–3012

Buller AHR (1933) The translocation of protoplasm through septate mycelium of certain Pyrenomycetes, Discomycetes and Hymenomycetes. In: Researches on fungi, vol 5. Longmans, Green & Co, London, pp 75–167

Camp RR (1977) Association of microbodies, Woronin bodies, and septa in intercellular hyphae of *Cymadothea trifolii*. Can J Bot 55:1856–1859

Curach NC, Te'o VS, Gibbs MD, Bergquist PL, Nevalainen KM (2004) Isolation, characterization and expression of the *hex1* gene from *Trichoderma reesei*. Gene 331:133–140

Edgerton-Morgan H, Oakley BR (2012) gamma-Tubulin plays a key role in inactivating APC/C(Cdh1) at the G(1)-S boundary. J Cell Biol 198:785–791

Engh I, Würtz C, Witzel-Schlömp K, Zhang HY, Hoff B, Nowrousian M, Rottensteiner H, Kück U (2007) The WW domain protein PRO40 is required for fungal fertility and associates with Woronin bodies. Eukaryot Cell 6:831–843

Escaño CS, Juvvadi PR, Jin FJ, Takahashi T, Koyama Y, Yamashita S, Maruyama J, Kitamoto K (2009) Disruption of the *Aopex11-1* gene involved in peroxisome proliferation leads to impaired Woronin body formation in *Aspergillus oryzae*. Eukaryot Cell 8:296–305

Fiddy C, Trinci AP (1976) Mitosis, septation, branching and the duplication cycle in *Aspergillus nidulans*. J Gen Microbiol 97:169–184

Fleißner A, Glass NL (2007) SO, a protein involved in hyphal fusion in *Neurospora crassa*, localizes to septal plugs. Eukaryot Cell 6:84–94

Fleißner A, Sarkar S, Jacobson DJ, Roca MG, Read ND, Glass NL (2005) The *so* locus is required for vegetative cell fusion and postfertilization events in *Neurospora crassa*. Eukaryot Cell 4:920–930

Han P, Jin FJ, Maruyama J, Kitamoto K (2014) A large nonconserved region of the tethering protein Leashin is involved in regulating the position, movement, and function of Woronin bodies in *Aspergillus oryzae*. Eukaryot Cell 13:866–877

Jedd G (2011) Fungal evo-devo: organelles and multicellular complexity. Trends Cell Biol 21:12–19

Jedd G, Chua NH (2000) A new self-assembled peroxisomal vesicle required for efficient resealing of the plasma membrane. Nat Cell Biol 2:226–231

Jedd G, Pieuchot L (2012) Multiple modes for gatekeeping at fungal cell-to-cell channels. Mol Microbiol 86:1291–1294

Juvvadi PR, Maruyama J, Kitamoto K (2007) Phosphorylation of the *Aspergillus oryzae* Woronin body protein, AoHex1, by protein kinase C: evidence for its role in the multimerization and proper localization of the Woronin body protein. Biochem J 405:533–540

Keller GA, Krisans S, Gould SJ, Sommer JM, Wang CC, Schliebs W, Kunau W, Brody S, Subramani S (1991) Evolutionary conservation of a microbody targeting signal that targets proteins to peroxisomes, glyoxysomes, and glycosomes. J Cell Biol 114:893–904

Kim KK, Hung LW, Yokota H, Kim R, Kim SH (1998) Crystal structures of eukaryotic translation initiation factor 5A from *Methanococcus jannaschii* at 1.8 A resolution. Proc Natl Acad Sci USA 95:10419–10424

Kim KH, Willger SD, Park SW, Puttikamonkul S, Grahl N, Cho Y, Mukhopadhyay B, Cramer RA Jr, Lawrence CB (2009) TmpL, a transmembrane protein required for intracellular redox homeostasis and virulence in a plant and an animal fungal pathogen. PLoS Pathog 5:e1000653

Kubo Y, Fujihara N, Harata K, Neumann U, Robin GP, O'Connell R (2015) *Colletotrichum orbiculare FAM1* encodes a novel Woronin body-associated Pex22 peroxin required for appressorium-mediated plant infection. MBio 6:e01305–e01315

Lai J, Koh CH, Tjota M, Pieuchot L, Raman V, Chandrababu KB, Yang D, Wong L, Jedd G (2012) Intrinsically disordered proteins aggregate at fungal cell-to-cell channels and regulate intercellular connectivity. Proc Natl Acad Sci USA 109:15781–15786

Leonhardt Y, Kakoschke SC, Wagener J, Ebel F (2017) Lah is a transmembrane protein and requires Spa10 for stable positioning of Woronin bodies at the septal pore of *Aspergillus fumigatus*. Sci Rep 7:44179

Lew RR (2005) Mass flow and pressure-driven hyphal extension in *Neurospora crassa*. Microbiology 151:2685–2692

Li H, Oberhauser AF, Redick SD, Carrion-Vazquez M, Erickson HP, Fernandez JM (2001) Multiple conformations of PEVK proteins detected by single-molecule techniques. Proc Natl Acad Sci USA 98:10682–10686

Li L, Wang J, Zhang Z, Wang Y, Liu M, Jiang H, Chai R, Mao X, Qiu H, Liu F, Sun G (2014) MoPex19, which is essential for maintenance of peroxisomal structure and Woronin bodies, is required for metabolism and development in the rice blast fungus. PLoS One 9:e85252

Lichius A, Yáñez-Gutiérrez ME, Read ND, Castro-Longoria E (2012) Comparative live-cell imaging analyses of SPA-2, BUD-6 and BNI-1 in *Neurospora crassa* reveal novel features of the filamentous fungal polarisome. PLoS One 7:e30372

Liu F, Ng SK, Lu Y, Low W, Lai J, Jedd G (2008) Making two organelles from one: Woronin body biogenesis by peroxisomal protein sorting. J Cell Biol 180:325–339

Liu F, Lu Y, Pieuchot L, Dhavale T, Jedd G (2011) Import oligomers induce positive feedback to promote peroxisome differentiation and control organelle abundance. Dev Cell 21:457–468

Managadze D, Würtz C, Sichting M, Niehaus G, Veenhuis M, Rottensteiner H (2007) The peroxin PEX14 of *Neurospora crassa* is essential for the biogenesis of both glyoxysomes and Woronin bodies. Traffic 8:687–701

Markham P (1994) Occlusions of septal pores in filamentous fungi. Mycol Res 98:1089–1106

Markham P, Collinge AJ (1987) Woronin bodies of filamentous fungi. FEMS Microbiol Rev 46:1–11

Maruyama J, Kitamoto K (2007) Differential distribution of the endoplasmic reticulum network in filamentous fungi. FEMS Microbiol Lett 272:1–7

Maruyama J, Juvvadi PR, Ishi K, Kitamoto K (2005) Three-dimensional image analysis of plugging at the septal pore by Woronin body during hypotonic shock inducing hyphal tip bursting in the filamentous fungus *Aspergillus oryzae*. Biochem Biophys Res Commun 331:1081–1088

Maruyama J, Kikuchi S, Kitamoto K (2006) Differential distribution of the endoplasmic reticulum network as visualized by the BipA-EGFP fusion protein in hyphal compartments across the septum of the filamentous fungus, *Aspergillus oryzae*. Fungal Genet Biol 43:642–654

Maruyama J, Escaño CS, Kitamoto K (2010) AoSO protein accumulates at the septal pore in response to various stresses in the filamentous fungus *Aspergillus oryzae*. Biochem Biophys Res Commun 391:868–873

Mason PJ, Crosse R (1975) Crystalline inclusions in hyphae of the glaucus group of Aspergilli. Trans Br Mycol Soc 65:129–134

McKeen WE (1971) Woronin bodies in *Erysiphe graminis* DC. Can J Microbiol 17:1557–1560

Momany M, Richardson EA, Van Sickle C, Jedd G (2002) Mapping Woronin body position in *Aspergillus nidulans*. Mycologia 94:260–266

Müller WH, Montijn RC, Humbel BM, van Aelst AC, Boon EJ, van der Krift TP, Boekhout T (1998) Structural differences between two types of basidiomycete septal pore caps. Microbiology 144:1721–1730

Nave R, Fürst DO, Weber K (1989) Visualization of the polarity of isolated titin molecules: a single globular head on a long thin rod as the M band anchoring domain? J Cell Biol 109:2177–2187

Nayak T, Edgerton-Morgan H, Horio T, Xiong Y, De Souza CP, Osmani SA, Oakley BR (2010) Gamma-tubulin regulates the anaphase-promoting complex/cyclosome during interphase. J Cell Biol 190:317–330

Ng SK, Liu F, Lai J, Low W, Jedd G (2009) A tether for Woronin body inheritance is associated with evolutionary variation in organelle positioning. PLoS Genet 5:e1000521

Oldfield CJ, Dunker AK (2014) Intrinsically disordered proteins and intrinsically disordered protein regions. Annu Rev Biochem 83:553–584

Peat TS, Newman J, Waldo GS, Berendzen J, Terwilliger TC (1998) Structure of translation initiation factor 5A from *Pyrobaculum aerophilum* at 1.75 A resolution. Structure 6:1207–1214

Plamann M (2009) Cytoplasmic streaming in *Neurospora*: disperse the plug to increase the flow? PLoS Genet 5:e1000526

Ryan FJ, Beadle GW, Tatum EL (1943) The tube method of measuring the growth rate of Neurospora. Am J Bot 30:784–799

Shen KF, Osmani AH, Govindaraghavan M, Osmani SA (2014) Mitotic regulation of fungal cell-to-cell connectivity through septal pores involves the NIMA kinase. Mol Biol Cell 25:763–775

Son M, Lee KM, Yu J, Kang M, Park JM, Kwon SJ, Kim KH (2013) The HEX1 gene of *Fusarium graminearum* is required for fungal asexual reproduction and pathogenesis and for efficient viral RNA accumulation of *Fusarium graminearum* virus 1. J Virol 87:10356–10367

Soundararajan S, Jedd G, Li X, Ramos-Pamploña M, Chua NH, Naqvi NI (2004) Woronin body function in *Magnaporthe grisea* is essential for efficient pathogenesis and for survival during nitrogen starvation stress. Plant Cell 6:1564–1574

Steinberg G, Schuster M, Hacker C, Kilaru S, Correia A (2017a) ATP prevents Woronin bodies from sealing septal pores in unwounded cells of the fungus *Zymoseptoria tritici*. Cell Microbiol 19:e12764

Steinberg G, Harmer NJ, Schuster M, Kilaru S (2017b) Woronin body-based sealing of septal pores. Fungal Genet Biol 109:53–55

Tegelaar M, Wösten HAB (2017) Functional distinction of hyphal compartments. Sci Rep 7:6039

Teichert I, Steffens EK, Schnaß N, Fränzel B, Krisp C, Wolters DA, Kück U (2014) PRO40 is a scaffold protein of the cell wall integrity pathway, linking the MAP kinase module to the upstream activator protein kinase C. PLoS Genet 10:e1004582

Tenney K, Hunt I, Sweigard J, Pounder JI, McClain C, Bowman EJ, Bowman BJ (2000) Hex-1, a gene unique to filamentous fungi, encodes the major protein of the Woronin body and functions as a plug for septal pores. Fungal Genet Biol 31:205–217

Tey WK, North AJ, Reyes JL, Lu YF, Jedd G (2005) Polarized gene expression determines Woronin body formation at the leading edge of the fungal colony. Mol Biol Cell 16:2651–2659

Trinci APJ, Collinge AJ (1974) Occlusion of the septal pores of damaged hyphae of *Neurospora crassa* by hexagonal crystals. Protoplasma 80:57–67

Wergin WP (1973) Development of Woronin bodies from microbodies in *Fusarium oxysporum* f. sp. *lycopersici*. Protoplasma 76:249–260

Woronin M (1864) Zur Entwicklungsgeschichte der Ascobolus pulcherrimus Cr und eigiger Pezizen. Abh Senkenb Naturforsch Ges 5:333–344

Wright PE, Dyson HJ (2009) Linking folding and binding. Curr Opin Struct Biol 19:31–38

Würtz C, Schliebs W, Erdmann R, Rottensteiner H (2008) Dynamin-like protein-dependent formation of Woronin bodies in *Saccharomyces cerevisiae* upon heterologous expression of a single protein. FEBS J 275:2932–2941

Yuan P, Jedd G, Kumaran D, Swaminathan S, Shio H, Hewitt D, Chua NH, Swaminathan K (2003) A HEX-1 crystal lattice required for Woronin body function in *Neurospora crassa*. Nat Struct Biol (4):264–270

Zekert N, Veith D, Fischer R (2010) Interaction of the *Aspergillus nidulans* microtubule-organizing center (MTOC) component ApsB with gamma-tubulin and evidence for a role of a subclass of peroxisomes in the formation of septal MTOCs. Eukaryot Cell 9:795–805

Septum Formation and Cytokinesis in Ascomycete Fungi

Stephan Seiler[1,2], Yvonne Heilig[1]

CONTENTS

[1] Institute for Biology II – Molecular Plant Physiology, Albert-Ludwigs University Freiburg, Freiburg, Germany; e-mail: Stephan.Seiler@biologie.uni-freiburg.de;
[2] Freiburg Institute for Advanced Studies (FRIAS), Albert-Ludwigs University Freiburg, Freiburg, Germany

I. Introduction

Cell division is a fundamental cellular process that is essential for proliferation of unicellular as well as multicellular organisms. It reflects the final stage of the cell cycle, during which a cell is physically divided into two daughter cells that contain a full set of chromosomes and other cellular organelles. Cytokinesis can be divided into several general steps that apply to most eukaryotic cells (Barr and Gruneberg 2007; Eggert et al. 2006): (1) the selection of the future **cell division plane** based on spatial as well as temporal cues, (2) the **assembly of a cortical actomyosin ring (CAR)** at this site, and (3) its **constriction coupled with membrane invagination.** In general, the formation of the CAR and its subsequent constriction is tightly coupled to the cell cycle to ensure that cell separation does not occur prior to chromosome segregation. (4) The formation of an extracellular cell wall, the septum, composed of glucans, chitin, and other polysaccharides in fungi further requires coordination of CAR constriction with secretion of cell wall biosynthetic and remodeling enzymes to build the extracellular septum. (5) This **primary septum** is covered by additional layers of cell wall material that form the **secondary septum** and is finally degraded by secreted hydrolytic enzymes in the unicellular yeasts and during sexual development of filament-forming molds to allow detachment of the two cells.

Several overviews have recently summarized the mechanistic principles underlying cell polarization and division in budding yeast and fission yeast, two unicellular fungi that consis-

Biology of the Fungal Cell, 3rd Edition
The Mycota VIII
D. Hoffmeister, M. Gressler (Eds.)

tently serve as conceptual framework for the analysis of fungal as well as higher eukaryotic cell biology (Pollard and Wu 2010; Bi and Park 2012; Weiss 2012; Martin and Arkowitz 2014). However, the vast majority of fungi forms mycelial colonies that consist of networks of branched hyphe. These cells grow by tip growth and are compartmentalized by septum that partition cellular environments within the hypha. In contrast to the unicellular yeasts, not every nuclear division triggers the formation of a new septum in mycelial fungi, and thus hyphal compartments are generally multinucleate. Moreover, a small septal pore is retained in higher fungi to enable intercellular communication and transport of cytoplasm and organelles between adjacent hyphal compartments. This controlled segmentation of hyphal units through septal cross walls in a multicellular mycelium is the basis for the morphological complexity achieved by the fungi during vegetative growth, differentiation, and infection processes and is thus a prerequisite for the evolutionary success of the fungal kingdom (Gull 1978; Pringle and Taylor 2002; Blackwell 2011).

The phylum *Ascomycota* comprises three major subphyla: ***Taphrinomycotina***, ***Saccharomycotina***, and ***Pezizomycotina*** (McLaughlin et al. 2009; Stajich et al. 2009). The subphylum of the *Pezizomycotina* contains over 90% of the *Ascomycota* species. Almost all species of this clade generate multinuclear hyphae that are compartmentalized by septa and include the model molds *Aspergillus nidulans* and *Neurospora crassa*. The *Saccharomycotina* contain the industrial yeasts and parasitic *Canidida spec.* species, dimorphic fungi that can switch between yeast and mycelial states (Sudbery 2011). Most members of the *Saccharomycotina* are unicellular, but this group also includes filamentous forms, such as *Ashbya gossypii*. This species is very closely related to *Saccharomyces cerevisiae*, and more than 90% of the genes are highly conserved in the two fungi (Dietrich et al. 2004; Wendland and Walther 2005), but the two species have developed dramatically different growth forms: constitutive multinucleate tip growth in *A. gossypii* versus unicellular growth in *S. cerevisiae*. The third subphylum includes the *Schizosaccharomycetes* and other early diverging lineages and is primarily represented by *Schizosacchomyces pombe*. The monophyletic origin of this subphylum is still under debate, but most recent data support the monophyly of the taxon (James et al. 2006). Both filaments and yeasts are found in this subphylum, suggesting that both morphologies are ancestral in *Ascomycota*.

Despite the importance of septum for hyphae, proliferation, and differentiation in filamentous fungi, our understanding of septum formation and its regulation is highly fragmentary (Harris 2001; Seiler and Justa-Schuch 2010; Mourino-Perez and Riquelme 2013). In this review, we will focus on recent progress that confirms the use of conserved molecular modules during cell division in the unicellular yeasts and during septum formation in the filamentous fungi. However, it is also becoming apparent that proper placement and regulated formation and function of septa in the different phylogenetic groups of *Ascomycota* requires significant rewiring and species-specific adaptation of these conserved modules (Gu et al. 2015).

II. Spatial Cues: Mechanisms Specifying the Position of the Division Plane

A. Positioning the Cell Division Plane in Unicellular Yeasts

The regulatory pathways that control the spatial aspects to place the future cell division plane are poorly conserved among different eukaryotic organisms despite the high importance of a tight coordination of cytokinesis with chromosome and organelle segregation. For example, the two model yeasts *S. cerevisiae* and *S. pombe* have developed fundamentally distinct mechanisms to control the spatial aspects for placing the future cell division plane (Fig. 1a, b). The bud-site selection system of budding yeast relies on cortical cues of the previous cell division cycle in order to redirect the dividing nucleus to the bud neck, while the position

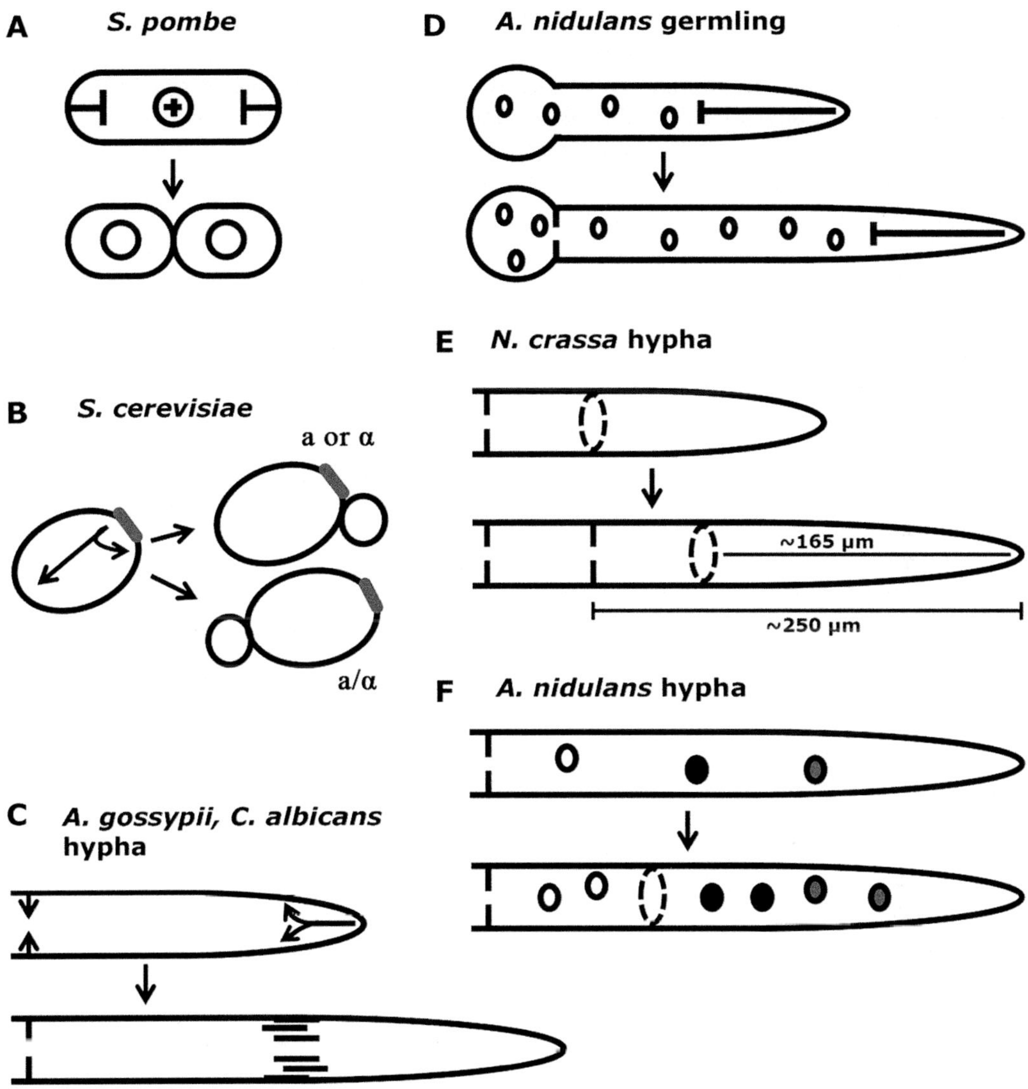

Fig. 1 Diverse types of spatial signals regulate septum placement and cell division plane. (**a**) The site of cell division in fission yeast is chosen based on a positive signal provided by the position of the pre-mitotic nucleus and negative signals from the two cell ends. (**b**) A transient cortical landmark at the bud scar functions as a spatial memory from one cell cycle to the next in budding yeast. *a* or *α* cells divide axially, such that dividing cells form their buds immediately adjacent to the site of the previous cell separation. *a/α* cells exhibit a bipolar budding pattern and place the new bud either proximal or distal to the site of the previous cell division. (**c**) Cortical cues deposited at the hyphal tip may trigger septum position in the filament-forming *Saccharomycotina* species *A. gossypii* and *C. albicans*. The cue for landmark deposition at the tip has been proposed to be reduced tip elongation as a consequence of septum formation in the subapical region of the hypha. (**d**) A tip-high inhibitory gradient has been proposed in *A. nidulans* to restrict septum formation in growing germlings until a certain cell size is reached. (**e**) A similar tip-high inhibitory signal has been proposed for mature *N. crassa* hyphae, where the new septum is placed when the apical cell reaches a critical size of approximately 250 µm. (**f**) Unlike *S. pombe*, where the position of the pre-mitotic nucleus serves as positive signal, the majority of septa in *A. nidulans* are formed at a position corresponding to a location in between two pairs of previously mitotic nuclei

of the fission yeast nucleus itself specifies the future plane of cell division (Chang and Peter 2003). Both mechanisms are dissected at the molecular level, and the reader is referred to recent reviews for details (Laporte et al. 2010; Pollard and Wu 2010; Bi and Park 2012; Weiss 2012).

Briefly, **budding yeast cells** divide in two precise spatial patterns (Chant and Herskowitz 1991). *a* or *α* cells divide axially, such that dividing cells form their buds immediately adjacent to the site of the previous cell separation. *a/α* cells exhibit a bipolar budding pattern and place the new bud either proximal or distal to the site of the previous cell division. The axial budding pattern depends on a transient cortical landmark consisting of the **Bud3p-Bud4p complex** that binds to the transmembrane glycoprotein Axl2p (Gao et al. 2007; Kang et al. 2012; Wu et al. 2015). The localization of these proteins acts as a spatial memory from one cell cycle to the next. An Axl2p containing spot localizes to the cell cortex at the time of bud emergence and serves together with the encircling septin ring as a landmark to recruit Bud3p and Bud4p. This Bud3p-Bud4p-Axl2 complex matures at the mother-bud neck into a double ring, and one ring is passed to each daughter cell, marking the previous site of attachment. This localization of the landmark complex at the mother-bud neck depends on the integrity of the septin collar; thus the Bud3p-Bud4p complex rings are likely assembling through the direct interaction with the septins (Wloka et al. 2011; Eluere et al. 2012; Kang et al. 2013). In the next round of budding, these cortical marker proteins recruit a **Ras-related GTPase module** (consisting of the GTPase Rsr1p/Bud1p, its GTPase-activating protein (GAP) Bud2p, and guanine nucleotide exchange factor (GEF) Bud5p) to the future bud site, which organizes the morphogenetic machinery toward the site of growth. At this site a new septin ring and Axl2p spot is formed generating a cyclic pattern of interdependence between the septins and the axial landmark proteins throughout the cell cycle.

The **fission yeast** *S. pombe* grows by elongation at its two ends and divides by medial fission, generating two roughly equally sized daughter cells. In contrast to budding yeast, in which the division site is solely determined by cortical cues, the division site in fission yeast is chosen based on a positive signal provided by the position of the pre-mitotic nucleus and negative signals from the two cell ends. Microtubules control the positioning of the nucleus at the cell center through opposing pushing forces generated by the dynamic instability of the microtubule system (Daga and Chang 2005; Tran et al. 2001). Furthermore, the bipolar longitudinal orientation of the microtubule system marks the cell ends by transporting the Tea1-Tea4 polarity complex to the tips, where it is tethered to the cortex through the prenylated protein Mod5 (Snaith et al. 2005; Snaith and Sawin 2003). This complex recruits the Dyrk family kinase Pom1 and other, yet undefined factors, to the cell ends (Bähler et al. 1998; Tatebe et al. 2005). A key factor that integrates both types of spatial signals is the anillin-related but *S. pombe*-specific protein Mid1. Mid1 reads the nuclear localization by shuttling between the nucleus and the adjacent cell cortex, where it forms a series of ca. 50 cortical dots that later in the cell cycle assemble and recruit CAR components (Almonacid et al. 2009; Celton-Morizur et al. 2006; Padte et al. 2006). The restriction of these interphase nodes to the cell center depends on Pom1-dependent and other inhibitory signals that are generated from both cell poles (Celton-Morizur et al. 2006; Huang et al. 2007; Padte et al. 2006; Rincon et al. 2014). In addition, Pom1 also phosphorylates the F-BAR protein Cdc15, a central component of the CAR, to inhibit CAR assembly at cell ends (Ullal et al. 2015).

B. Septum Placement in Filamentous Ascomycete Fungi

In contrast to the in-depth understanding of spatial signals present in the two unicellular yeast models, we have only very limited data about the presence and nature of positional cues that regulate septum placement in **filamentous ascomycete fungi**. It has been proposed that cortical cues deposited at the hyphal tip may trigger septum position in the filament-forming *Saccharomycotina* species

Ashbya gossypii (Kaufmann and Philippsen 2009; Fig. 1c). Although attractive, this hypothesis is primarily based on the observation that cortical marker proteins homologous to the *S. cerevisiae* Bud3p-Bud4p-septin landmark complex and associated proteins, such as the F-BAR scaffold protein Hof1, a central organizer of the CAR in both yeast models, are deposited at the hyphal tip in *A. gossypii* and *Candida albicans* at the time when tip growth slows as a consequence of septum formation in the subapical region of the hypha (Sudbery 2001; Wightman et al. 2004; Knechtle et al. 2003; DeMay et al. 2009; Helfer and Gladfelter 2006; Gonzalez-Novo et al. 2008; Kaufmann and Philippsen 2009). Moreover, effective progression through the cell cycle seems to require repositioning of migrating nuclei to these preselected sites for efficient initiation of septation (Alberti-Segui et al. 2001; Finley and Berman 2005; Helfer and Gladfelter 2006; Finley et al. 2008). Nuclear position relative to the imprinted septation sites may thus be a consequence of morphogenetic markers placed at the incipient bud site in unicellular or the tip growth in filamentous species of the *Saccharomycotina* clade. Clearly, this hypothesis needs further experimental support, especially because the mechanism(s) by which the bud-septin complex is deposited at the tip to mark future septation sites has not been addressed at the molecular level.

No changes in the rate of tip extension are observed during mitosis, septum formation, and branch initiation in ***A. nidulans*** and *N. crassa* hyphae (Horio and Oakley 2005; Jackson 2001; Riquelme and Bartnicki-Garcia 2004; Riquelme et al. 2003; Sampson et al. 2003). Thus, no growth rate-dependent septum to tip signal can exist in these species of the *Pezizomycotina* clade to mark future septation sites analogous to that proposed for the filamentous *Saccharomycotina* species. An alternative explanation may be an inhibitory gradient originating from the tip that overrides positive yet still undefined signals generated from the nuclei. Such a tip-high inhibitory gradient has been proposed in *A. nidulans* to restrict septum formation in growing germlings until a certain cell size is reached (Kaminskyj 2000; Wolkow et al. 1996; Harris 2001) (Fig. 1d). Although the nature of this proposed gradient is currently unknown, homologs of the *S. pombe* Tea1-Tea4 system may represent an attractive possibility (Fischer et al. 2008; Higashitsuji et al. 2009; Konzack et al. 2005; Takeshita et al. 2008, 2014; Takeshita and Fischer 2011). *A. nidulans* TeaA and TeaR, functional homologs of the *S. pombe* cell-end marker Tea1 and its membrane anchor Mod5, localize in an interdependent manner to the hyphal tip and colocalize there with the formin SepA (Takeshita et al. 2008). Deletion of either TeaA or TeaR results in wavy and meandering growth, indicating that the apical localization of these cell-end markers is required for stabilizing the axis of growth polarity. Interestingly, TeaC, the homolog of *S. pombe* Tea4, localizes to septa in addition to its presence at the hyphal tip (Higashitsuji et al. 2009). Overexpression of TeaC does not affect apical tip extension rates but represses septation and generates almost aseptate strains, while deletion of *teaC* results in increased septation, consistent with the hypothesis of positioning factors that inhibit septum formation in the tip region to regulate compartment size in growing hyphae. This view is consistent with recent data obtained for ***N. crassa*** that also support some kind of size-sensing mechanism for septum placement (Delgado-Alvarez et al. 2014; Fig. 1e). When mature *N. crassa* tip cells reach a critical size of ca. 250 μm, a new septum is initiated approximately 165 μm distal of the tip. However, unlike *S. pombe*, where the position of the pre-mitotic nucleus serves as additional positive signal, the majority of septa in *A. nidulans* are formed at a position corresponding to a location in between two pairs of previously mitotic nuclei (Fig. 1f; Shen et al. 2014), and the precise involvement of nuclear behavior in septum placement remains unclear (also see Sect. III-C).

C. Signal Integration by Anillin-Type Landmark Proteins

Despite the poor conservation of spatial cues, **anillin-type landmark proteins** coordinate these spatial signals and are thus critical for organizing the future site of cell division in all fungi that have been analyzed to date. Also, vertebrate anillin is among the first proteins that are recruited to the cleavage site of dividing cells (Cabernard et al. 2010; Piekny and Glotzer 2008), suggesting conserved functions of this class of poorly defined proteins in orchestrating cell division plate. In animals, anillin functions as scaffold for the **Rho GTPase RhoA** and its regulators ECT2/Pbl and RacGAP50C at the cleavage furrow (D'Avino et al. 2008; Gregory et al. 2008; Hickson and O'Farrell 2008). Also Bud4p and the fission yeast homolog, the Mid1-related protein Mid2, interact with specific Rho GTPase modules and are recruited to the incipient separation site in a septin-dependent manner. In budding yeast, the **axial landmark Bud3p, Bud4p, and Axl2p** assemble into a protein complex at mitosis. Septin filaments first recruit Bud4p and Bud3p, which interact through their C-termini, to the bud neck (Gao et al. 2007; Kang et al. 2012, 2013; Wu et al. 2015). Bud3p has weak GEF activity toward the Rho GTPase Cdc42p and can activate Cdc42p in vivo (Kang et al. 2014), which may be one important aspect for organizing the morphogenetic machinery. In fission yeast septin rings are involved primarily in cell-cell separation after the septum has formed. The fission yeast anillin Mid2 localizes as a ring in the middle of the cell after anaphase in a septin- and actin-dependent manner and influences septin ring organization (Tasto et al. 2003; Berlin et al. 2003). The GEF Gef3 interacts with $Rho3^{GTP}$ in vitro and functions as activator of Rho4 in vivo. Gef3 colocalizes and physically interacts with septins and Mid2 and requires septins and Mid2 for its localization (Munoz et al. 2014; Wang et al. 2015). Together these data support that Gef3 interacts with the septin complex and activates one or several Rho GTPases as a Rho GEF for septation in fission yeast. Similar modules, consisting of the Rho GTPase Rho4/RHO-4, its GEF Bud3/BUD-3, and the anillin Bud4/BUD-4, are required for septum formation in the filamentous ascomycetes *A. nidulans* and *N. crassa* (Justa-Schuch et al. 2010; Si et al. 2010, 2012; Rasmussen and Glass 2005). Thus, a general function of anillins may be the organization of a Rho GTPase-GAP-GEF module at the future cytokinetic site (D'Avino 2009; Zhang and Maddox 2010; Seiler and Justa-Schuch 2010).

The septins in budding yeast bud4Δ cells fail to form a double ring during cytokinesis, while overexpression of Bud4p causes extra septin structures (Wloka et al. 2011; Eluere et al. 2012; Kang et al. 2013), suggesting a positive feedback in the organization of septin collar and axial landmark complex. A positive feedback for the cortical recruitment of the anillin-RHO-4 GTPase module was also described for *N. crassa* (Justa-Schuch et al. 2010), where localization of BUD-3 depends on BUD-4, whose localizations in turn lead to the recruitment and activation of RHO-4. However, the stable cortex association of BUD-3 and BUD-4 also requires RHO-4. This feedback also allows the stable accumulation of the BUD-3-BUD-4-RHO-4 complex at presumptive septation sites prior to the initiation of septum formation.

Recent structural data of animal anillin and *S. pombe* Mid1 defined several functional regions of these two proteins despite their poor conservation at the sequence level (Fig. 2). The N-terminal regions of both proteins bind to multiple components of the CAR machinery, and Mid1 and animal anillin have functional exchangeable N-terminal domains (D'Avino 2009; Piekny and Maddox 2010; Watanabe et al. 2010; El Amine et al. 2013; Sun et al. 2015). This region can bind active myosin and can bundle actin filaments (Field and Alberts 1995; Straight et al. 2003). The central region of anillin carries an anillin-homology domain, which contains a coiled-coil region that—in animal anillin—harbors a well-defined RhoA-binding domain, followed by a membrane-binding C2 domain that forms β-sandwich structure. Moreover, the Mid1 C2 domain allows dimerization. The C-terminus of anillin possesses a PH domain that binds septins and phospholipids (Liu and Young 2012; Oegema et al. 2000; Sun et al. 2015). Although both animal anillin and Mid1 attach to the plasma membrane via the same core elements of the C2 domain and the conserved PH domains, the PH domain of Mid1 is not essential for membrane anchorage (Sun et al. 2015; Lee and Wu 2012; Paoletti and Chang 2000; Liu and Young 2012).

Most fungal anillin-like proteins cluster phylogenetically with *S. pombe* Mid2, which promotes cell separation as a late-acting septin-dependent scaffold, but does not influence cell

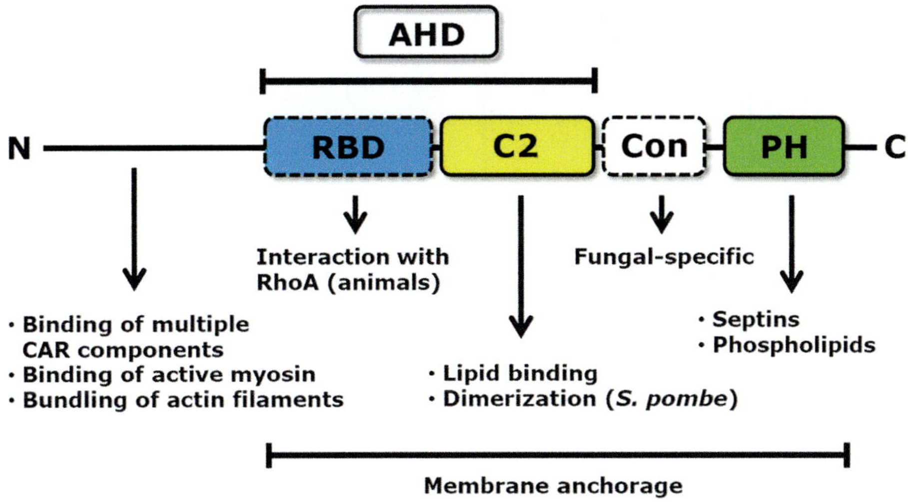

Fig. 2 Schematic representation of functional regions of fungal and animal anillins. *AHD* anillin-homology domain, *RBD* RhoA-binding domain, *C2* membrane-binding module, *CON* connector domain, *PH* Pleckstrin homology domain. For detailed explanation, refer to the text

division plate positioning. Mid1 and Mid2 arouse by gene duplication followed by functional divergence (a duplication that is specific for the genus *Schizosaccharomyces*), suggesting that the ancestral and conserved function of fungal anillin-related proteins is to scaffold CAR complexes in a septin-dependent manner (Gu and Oliferenko 2015; Gu et al. 2015; Moseley 2015). This is also supported by the recent finding that Mid1 of the close relative *Schizosaccharomyces japonicus* serves as an interphase cortical anchor for type II myosin, and the medial assembly of the CAR in mitotic *S. japonicus* cells relies on the cortical anchor protein Cdc15 instead of Mid1, which is regulated by a tip-localized Pom1 gradient (Gu and Oliferenko 2015; Gu et al. 2015; Moseley 2015).

III. Temporal Cue(s): Coordination of Cell Cycle Progression and Septation Initiation

A. Coordination of Mitotic Exit and Initiation of Septation in Unicellular Yeasts

The localization of the anillin-related landmark proteins Mid1 and Bud4p provides the spatial cue for septum placement in the two model yeasts. In addition, a GTPase-coupled kinase cascade—known as **mitotic exit network (MEN)** and **septation initiation network (SIN)** in budding and fission yeast, respectively—mediates the strict temporal coordination of cell cycle progression and cell division in the two unicellular fungi (Fig. 3a, b; Meitinger et al. 2012, Johnson et al. 2012). At the end of mitosis, the **spindle pole body (SPB)**-associated ras superfamily GTPase Spg1 is activated by the polo kinase Plo1 **in fission yeast.** This is achieved by phosphorylation, and thereby inhibition of its bipartite GTPase-activating protein (GAP) Cdc16-Byr4 results in recruitment of the STE kinase Cdc7 to activated Spg1 at the SPBs. Two additional SPB-associated kinases Sid1 and Sid2, with their respective regulatory subunits Cdc14 and Mob1, are part of the **SIN.** Localization studies suggest a hierarchical order of Cdc7-Sid1-Sid2, although biochemical evidence for a linear cascade is lacking (Johnson et al. 2012). Phosphorylation of the Cdc14 phosphatase Clp1 by the nuclear Dbf2p-related (NDR) effector kinase Sid2 promotes mitotic exit by counteracting the function of the cyclin kinase Cdc2 (Chen et al. 2008a). However, the SIN is not required for controlling mitotic exit, and SIN inactivation generates multinucleated

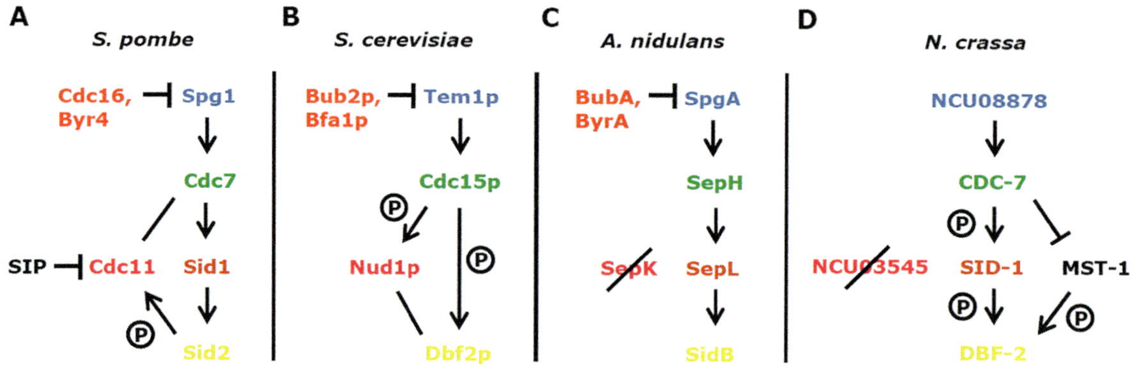

Fig. 3 Composition and regulation of the septation initiation network. (**a**) A GTPase-coupled, tripartite kinase cascade regulates initiation of septation in *S. pombe*. SPB-association of the effector kinase Sid2 depends on the two upstream kinases Sid1 and Cdc7, yet biochemical proof of a stepwise phosphorylation and activation of the three kinases is lacking. Positive feedback is provided by Sid2-dependent phosphorylation of the scaffold Cdc11, which is counteracted by the septation-inhibitory phosphatase complex SIP. (**b**) The related mitotic exit network (MEN) of budding yeast lacks a homolog of the fission yeast kinase Sid1, and the effector kinase Dbf2p is directly phosphorylated by the Cdc7 homolog Cdc15p. Recruitment of Dbf2p to SPBs and MEN activation also requires phosphorylation of the MEN scaffold Nud1p by Cdc15p. (**c**) A tripartite kinase cascade homologous to the fission yeast SIN also regulates septum formation in *A. nidulans*. However, SPB association of the SIN kinases through the scaffold SepK is not critical for septum formation, and SidB activation likely occurs in the cytosol. (**d**) The *N. crassa* SIN acts as cascade of three kinases, in which CDC-7 promotes the activity of SID-1, which in turn activates DBF-2 through hydrophobic motif phosphorylation. An additional kinase, MST-1, acts in parallel to SID-1 and is negatively regulated by interaction with CDC-7. This may allow fine-tuning of the SIN by generating an incoherent type 4 feedforward loop through the two, parallel functioning, but oppositely regulated CDC-7/SID-1 and CDC-7/MST-1 complexes that together control DBF-2 activity. SPB-association of these SIN components is not required for septum formation

cells, while overexpression of positive-acting SIN components causes the formation of multiple septum (Fankhauser and Simanis 1994; Ohkura et al. 1995; Schmidt et al. 1997). In both scenarios cell division is uncoupled from nuclear division. The assembly and subsequent constriction of the CAR that triggers septum formation requires the relocation of the active Sid2-Mob1 kinase complex, yet not the other SIN components, from the SPBs to the cell cortex (Roberts-Galbraith and Gould 2008; Hachet and Simanis 2008). The mechanisms that target the Sid2-Mob1 complex to the cell cortex are yet elusive.

The related **mitotic exit network (MEN) of budding yeast** has a similar composition and functions in an analogous manner (Meitinger et al. 2012). However, several significant differences exist (Fig. 3a, b). First, the MEN lacks a homolog of the fission yeast kinase Sid1, and the effector kinase Dbf2p is directly phosphorylated by the Cdc7 homolog Cdc15p (Mah et al. 2001). The Dbf2p phosphosites identified in this study correspond to sites highly conserved in other fungal NDR kinases (Hou et al. 2004; Jansen et al. 2006; Maerz et al. 2012), suggesting that Dbf2p activation involves activation segment autophosphorylation and phosphorylation of its C-terminal hydrophobic motif through Cdc15p. Second, recruitment of the Dbf2p-Mob1p complex to SPBs (and thus MEN activation) requires phosphorylation of the MEN scaffold protein Nud1p by upstream kinase Cdc15p (Mah et al. 2001; Rock et al. 2013). This differs from the positive feedback mechanism of the fission yeast SIN assembly at the SPB, in which the scaffold Cdc11 is phosphorylated by the effector kinase Sid2 to promote enhanced interaction with the upstream kinase Cdc7 (Feoktistova et al. 2012). Third, budding yeast MEN strictly controls mitotic exit, and consequently *MEN* mutants arrest at a late mitotic state with a CAR that has formed, but is unable to constrict (Yoshida et al. 2006), while *S. pombe SIN* mutants produce multinucleate cells.

B. The Septation Initiation Network in the *Pezizomycotina* Fungi

The MEN also controls mitotic spindle orientation and nuclear segregation in yeast cells of ***C. albicans***. Depletion of the MEN effector kinase Dbf2 results in growth arrest after the assembly of the actomyosin ring, indicating that **MEN activity is also essential for CAR contraction**, but not for CAR assembly (Gonzalez-Novo et al. 2009). Consequently septum formation is also inhibited in *C. albicans* filaments. As in yeast cells, Calcofluor White-stainable material still accumulates at presumptive septation sites, suggesting that CAR assembly is still possible and constriction is blocked, but this has not been addressed experimentally. However, not all *C. albicans MEN* mutant share identical defects, arguing against a simple and linear signal cascade and supporting distinct functions of individual MEN components. For example, unlike Tem1p-depleted *S. cerevisiae* cells, which arrest as large-budded cells as any other *MEN* mutant in budding yeast, GTPase-depleted *C. albicans* forms filaments that originate from large-budded yeast cells, suggesting that the GTPase Tem1 has an important, MEN-independent function for filament induction (Milne et al. 2014). The hyphae generated in a Tem1-depleted strain are binucleated and arrested in telophase with an elongated spindle, indicating that the MEN operates normally in these filaments and that MEN deficiency blocks mitotic exit after one round of nuclear division.

In contrast, the ***A. gossypii*** **MEN** has been suggested to function as linear GTPase-coupled bipartite kinase cascade, an assumption that was based on common mutant defects of deleting any MEN component (Finlayson et al. 2011). However, in contrast to *S. cerevisiae* or *C. albicans*, where the MEN controls mitotic exit at the anaphase stage, *A. gossypii MEN* mutants are enriched for metaphase nuclei, and thus the MEN seems to function earlier in the cell cycle of this organism. Also, cell cycle control is not absolute, and the multinuclear status of the hyphal compartments is maintained. As in *C. albicans*, the *A. gossypii* MEN is also essential for septum formation, and *MEN* mutants form aseptate hyphae (Finlayson et al. 2011).

In summary, a common denominator for the MEN in the three *Saccharomycotina* species seems to be a central role in CAR constriction but not CAR assembly. The regulation of the MEN and a mechanistic basis for its importance during septum formation in the filamentous context has not been addressed.

The **SIN in the *Pezizomycotina*** models *A. nidulans* and *N. crassa* functions as three-leveled kinase cascade that is activated by an upstream GTPase module as described for fission yeast. Deletion of any positive-acting SIN component in either fungus results in aseptate strains with no CAR formed, while no involvement in cell cycle control has been described (Bruno et al. 2001; Kim et al. 2006, 2009; Heilig et al. 2013; Maerz et al. 2009). All SIN components localize to the SPBs and—with the exception of the scaffold proteins and the upstream kinase—to constricting septa (Heilig et al. 2013; Kim et al. 2009). The *N. crassa* SIN components accumulate at the cell cortex several minutes prior to initiation of septum constriction, arguing for their involvement in CAR assembly, while CDC-7 is recruited to the forming septum only after septum constriction had started (Heilig et al. 2013, 2014). Similarly, the SidB-MobA module of *A. nidulans* associates with constricting septa (Kim et al. 2006, 2009), while SepH does neither associate with forming nor mature septum (De Souza et al. 2014).

Although the function of the SIN as linear acting GTPase-coupled kinase device has originally been assumed for both fungi (e.g., Harris 2001; Seiler and Justa-Schuch 2010), recent data suggest a more complicated network of kinases (Fig. 3c, d). The ***N. crassa*** **SIN** acts as cascade of three kinases, in which CDC-7 promotes the activity of SID-1, which in turn activates DBF-2 through hydrophobic motif phosphorylation (Heilig et al. 2013). However, an additional kinase, MST-1, acts in parallel to SID-1 and is negatively regulated by interaction with CDC-7 (Heilig et al. 2014). A possible function of MST-1 in fine-tuning the SIN may be achieved in the generation of an incoherent type 4 feedforward loop through the two, parallel functioning, but oppositely regulated CDC-7/SID-1 and CDC-7/MST-1 complexes that together control DBF-2 activity. This rarely found regulatory motif may allow for adaptive and thus robust SIN activity (Rodrigo and Elena 2011). This hypothesis is supported by the phenotypic characteristics of *N. crassa SIN* mutants, which revealed less severe septation defects of *sid-1* and *mst-1* mutants compared to strains deficient for *cdc-7* or *dbf-2* (Maerz et al. 2009; Heilig et al. 2013, 2014). The biochemical data obtained in this study also imply that the CDC-7/MST-1 complex may function independently of the GTPase SPG-1 (Heilig et al. 2014), a speculation that requires further exploration.

Interestingly, MST-1 is also essential for sexual development (Heilig et al. 2014), and mutants in the *Sordaria macrospora* homolog SmSTK24 was recently found to interact with the STRIPAK (striatin-associated phosphatase and kinase) complex (Bloemendal et al. 2012; Frey et al. 2015). Although the *N. crassa* STRIPAK complex does not localize to septa and has primarily been associated with nuclear-cytoplasmic distribution of the cell wall integrity MAP kinase MAK-1 (Dettmann et al. 2013), several *S. macrospora STRIPAK* mutants form aseptate ascogonia (although vegetative septation is normal (Bloemendal et al. 2010). It is tempting to predict the involvement of the STRIPAK complex at an early stage of sexual development in the *Pezizomycotina* fungi. Intriguingly, the *S. pombe* STRIPAK complex has also been associated with regulating the SIN. This SIN-inhibitory phosphatase (SIP) complex counteracts the positive feedback generated by Sid2-dependent phosphorylation of Cdc11 by dephosphorylating the scaffold (Singh et al. 2011).

C. Coordination of Nuclear Behavior and Septum Formation in *Pezizomycotina* Fungi

CAR assembly and septum formation are clearly controlled through nuclear position and cell cycle progression in *A. nidulans* (Harris et al. 1994; Wolkow et al. 1996). This may potentially also apply to *N. crassa*, although the connection between nuclear cycle and septum positioning is blurred by its nuclear asynchrony (Plamann et al. 1994; Minke et al. 1999; Seiler and Justa-Schuch 2010). The recent finding that circadian rhythms can synchronize mitosis via the *N. crassa* homolog of the WEE-1 kinase (Hong et al. 2014) might provide a platform for a more detailed analysis. Septum placement was originally proposed to depend on the position of mitotic nuclei in *A. nidulans* (Momany and Hamer 1997; Bruno et al. 2001; Harris 2001), but detailed life imaging recently revealed that the majority of septa formed at a position corresponding to a location in between two pairs of previously mitotic nuclei (Shen et al. 2014; Fig. 1c). Moreover, septum formation still occurs when the microtubule cytoskeleton was destroyed, indicating that initiation and completion of septation does not require microtubules.

Although we have currently no mechanistic understanding concerning the coordination of nuclear behavior and septation, a number of candidate proteins/pathways have been analyzed, the most obvious being the SIN. Although all components of the SIN localize to SPBs in *A. nidulans* and *N. crassa*, no involvement in controlling cell cycle progression has been observed in either species during vegetative growth (Bruno et al. 2001; Kim et al. 2009; Heilig et al. 2013, 2014). Moreover, in contrast to the yeasts, where SPB association of the SIN/MEN components is essential for activation of the NDR effector kinase and the subsequent recruitment of the NDR kinase-Mob1 complex to the cell cortex for cytokinesis (Morrell et al. 2004; Rosenberg et al. 2006), SIN activation and septum formation in *A. nidulans* and *N. crassa* does not require SPB association (Kim et al. 2009; Heilig and Seiler, unpublished). This is consistent with the observation that the *A. nidulans* Polo kinase PlkA is not a central regulator of septum formation and has only a minor function in cell cycle control (Bachewich et al. 2005; Mogilevsky et al. 2012). Consequently, activation of the SIN may not be mediated through PlkA as established for the MEN/SIN through the budding and fission yeast Polo kinases Cdc5p and Plo1, respectively (Ohkura et al. 1995; Song et al. 2000). Finally, the two scaffold proteins SnaD and SepK (structural homologs of Sid4 and Cdc11 in *S. pombe*, respectively) that are necessary to anchor SIN components to the spindle pole body in *A. nidulans* are not critical for septum formation, and mutations in either scaffold only result in delayed septation (Kim et al. 2009). Thus, SPB association of the SIN might modulate temporal aspects of septum formation, but details of a mechanistic involvement of the SIN in coordinating nuclear behavior with septation during vegetative hyphal growth remain obscure.

Interestingly, it was recently shown that SepH concentrates in the basal region of the apical cells as cytosolic foci in addition to its association with mitotic SPB in *A. nidulans* (De Souza et al. 2014). Cytosolic SepH foci were particularly prominent in periods preceding septation and resulted in a preferential association of SepH with the SPB of mitotic nuclei, which were located distal from the tip. Moreover, SepH associated with SPB in a biphasic manner first during mitosis and again during the period of septum formation, suggesting that the predominant-basal activation of the SIN [if SPB-association can serve as marker for SIN activity as described for fission yeast; (Johnson et al. 2012)] might help restricting SIN activity to the basal region of the tip cell to promote asymmetric septation in *A. nidulans*. These cytosolic foci were also observed for the *N. crassa* homolog CDC-7 (Heilig et al. 2013, 2014) but not for any of the other SIN kinases or respective regulatory subunits in *N. crassa* (Heilig et al. 2013, 2014; Dettmann et al. 2012) or the SidB-MobA module in *A. nidulans* (Kim et al. 2006, 2009). CDC-7 kinase activity is not required for cytosolic clustering of the protein and its association with SPBs in *N. crassa*, and CDC-7 foci were more abundant when a kinase dead version of CDC-7 was localized (Heilig et al. 2014). No asymmetric localization of CDC-7 was noticed in *N. crassa*, but only mature hyphae were analyzed, in which fast cytosolic flow rates likely had mixed any potential gradient of CDC-7 clusters. A potential function of these cytosolic CDC-7 speckles might be suggested by the co-purification of the **microtubule-organizing center (MTOC)** component NCU12004 (homolog of *A. nidulans* ApsB and *S. pombe* Mto1) with CDC-7 yet with none of the other SIN components in *N. crassa* (Heilig et al. 2014). MT nucleation occurs primarily at centrosomes/SPBs, but more diverse types of cytosolic MTOC also exist. The mechanisms for the generation of this MTOC diversity are poorly understood, but it is essential for functional MT organization in fungi and higher eukaryotes. Mto1/ApsB is central for cytosolic MTOC assembly and function (Samejima et al. 2010; Lynch et al. 2014), and ApsB is recruited to the septal pore through a peroxisome-dependent pathway to form septum-associated MTOCs (Veith et al. 2005; Zekert et al. 2010). Intriguingly, CDC-7 localization around the mature septal pore in *N. crassa* (Heilig et al. 2014) displays striking similarities with the sepal pore localization of ApsB and γ-tubulin in *A. nidulans* (Zekert et al. 2010). Future analysis should therefore consider the speculation that these cytosolic clusters might represent non-SPB-associated MTOCs.

Other candidate proteins that are involved in coordinating nuclear behavior with septum biology in *A. nidulans* are the cyclin kinase $NimX^{Cdk1}$ and the NEK-family kinase NimA. Septum formation depends on a threshold level of activity of the $NimX^{Cdk1}$ (Harris 2001; Harris and Kraus 1998; Kraus and Harris 2001). How $NimX^{Cdk1}$ coordinates nuclear division with septum in *A. nidulans* remains unresolved. NimA functions as central regulator of multiple stage-specific aspects of mitosis. Key functions of NimA are chromatin condensation through histone H3 phosphorylation and the partial opening of the nuclear envelope though disassembly of nuclear pore complex (De Souza et al. 2000, 2004). NimA also has interphase-specific functions such as regulating microtubule dynamics and thus tip growth and was shown to localize at the tip growth and microtubule plus ends in a EB1-dependent manner (Govindaraghavan et al. 2014). Important in the context of septum formation is the recruitment of NimA to the cell cortex prior to the initiation of septum constriction (Shen et al. 2014). Intriguingly, the fission yeast NimA homolog Fin1 is activated for mitotic commitment by phosphorylation through the NDR kinase Sid2 (Grallert et al. 2012; not to be confused with Plo1-dependent activation of the SIN at mitotic exit). In G2 phase, the Sid2-Mob1 complex acts independently of other SIN components such as the scaffold Cdc11 to control the timing of mitotic entry, suggesting that Sid2 activation does not occur at the SPBs. Although highly speculative, activation and/or recruitment of

NimA by SidB to cortical sites (primed by unknown cues) might be involved in triggering initiation of septum constriction. Moreover, in contrast to other protein machinery required for septum formation in *A. nidulans* such as type II myosin MyoB (Hill et al. 2015), NimA is maintained at the septal pore after septum constriction has terminated (Shen et al. 2014). It is thus possible that NimA at this position controls the regulated closure of septal pores in a cell cycle-dependent manner in order to restrict leakage of nuclear proteins into subapical compartments during the parasynchronous mitoses of the nuclei present in the apical cell.

IV. Assembly and Regulation of the Contractile Actomyosin Ring (CAR)

A. CAR Assembly in Budding and Fission Yeast

Significant advances have allowed establishing the approximate spatiotemporal sequence of events during assembly and constriction of the CAR that drives cytokinesis in the two yeasts (Bi and Park 2012; Pollard and Wu 2010). In fission yeast, two independent but synergistic pathways driven by the anillin-related landmark protein Mid1 and the SIN control assembly of the CAR (Pollard and Wu 2010, Bathe and Chang 2010, Johnson et al. 2012; Fig. 4a). Interestingly, both pathways are controlled by the **Polo kinase Plo1.** This central cell cycle kinase is therefore exerting spatial control through regulating Mid1 localization and is also providing the temporal cue through regulation of SIN activity. Cortical, Mid1-containing dots recruit class II myosin, the formin Cdc12, and other components to form medial nodes that allow F-actin nucleation early in mitosis. Search-capture-pull-release interactions between myosin and f-actin from distinct nodes form a highly organized ring (Wu et al. 2006; Vavylonis et al. 2008; Laporte et al. 2011; Lee and Wu 2012; Saha and Pollard 2012). Also, actin filaments assembled by the formin Cdc12 at nonmedial sites are transported to the division site in a myosin-dependent manner and are integrated into the maturing ring (Huang

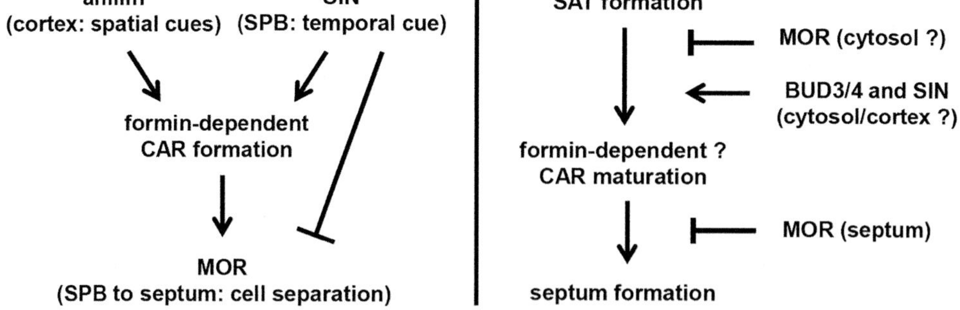

Fig. 4 Spatiotemporal sequence of events that control assembly and constriction of the CAR. (**a**) Two independent, but synergistic pathways driven by anillin-related landmark proteins and the SIN/MEN control assembly and/or constriction of the CAR in budding and fission yeast. SIN activity also inhibits the MOR network, which is required for timely cell separation. (**b**) Unknown cues initiate the formation of the septal actomyosin tangle SAT in the *Pezizomycotina* fungi. SAT initiation, CAR formation, and CAR constriction are regulated at various levels by the anillin-Rho GTPase complex and the interplay of the SIN and MOR kinase networks

et al. 2012). However, Mid1 nodes are not essential for CAR assembly, and *mid1* mutants produce misplaced septa in the anaphase. This indicates the presence of a second, SIN-dependent pathway that is responsible for CAR assembly in late mitosis (Huang et al. 2008; Hachet and Simanis 2008; Roberts-Galbraith and Gould 2008). Because *sin* mutants form only a transient CAR that is instable, these data suggest that the primary task of Mid1 is providing positional information, while the SIN is required for CAR maturation. Moreover, CAR constriction also requires SIN activity.

The maturation of the CAR from cortical nodes requires phosphorylation of several independent targets by the terminal SIN kinase Sid2. First, the recruitment of the F-BAR scaffold protein Cdc15 to the medial region of the cell is indirectly controlled by the SIN (Clifford et al. 2008; Hachet and Simanis 2008; Roberts-Galbraith et al. 2010). Interphase Cdc15 is phosphorylated at multiple sites that induces a closed conformation of the protein and inhibits its assembly at the division site. Activation of the Cdc14 phosphatase Clp1 by the SIN effector kinase Sid2 allows association of Clp1 with Mid1 and dephosphorylation of Cdc15. This induces an open conformation and oligomerization of Cdc15 that activates its scaffold function. Second, the SIN directly activates the formin Cdc12 through Sid-2-dependent phosphorylation of a novel Cdc12 domain that controls F-actin bundling and therefore a central function of the formin during CAR maturation (Bohnert et al. 2013). Importantly, this novel Cdc12 function should not be confused with the central function of formins in F-actin nucleation, which is controlled through dimerization of the FH2 domains (Xu et al. 2004). Although is it clear that the SIN is also important during septum constriction, the mechanism that triggers initiation of CAR constriction and the specific functions of the SIN during constriction are yet unknown (Johnson et al. 2012).

The molecular composition of the budding yeast CAR is very similar to that of fission yeast although the CAR components accumulate over longer time periods (Luo et al. 2004; Shannon and Li 2000; Wloka and Bi 2012; Balasubramanian et al. 2004; Bi et al. 1998; Lippincott and Li 1998). One important difference between the two yeasts is that the CAR assembles at the bud neck only after the septin collar has formed in response to the Bud3p-Bud4p-Axl2p landmark described earlier (Gladfelter et al. 2001; McMurray and Thorner 2009). As discussed before, the future septation site in the filament-forming *Saccharomycotina* species *A. gossypii* and *C. albicans* may be marked at the hyphal tip as response of the reduced growth rate triggered by subapical septum formation. This apical landmark consists of *Ashbya gossypii* Bud3 or the *C. albicans* Bud4 homolog Int1 and various septins in both fungi (Wendland 2003; DeMay et al. 2009; Gale et al. 2001). This allows recruitment of proteins required for CAR assembly, such as the F-BAR protein Hof1 (homolog of *S. pombe* Cdc15), type II myosin, and the IQGAP protein Cyk1 to form cortical filaments or “bars” within 5–10 μm of the hyphal tip, which are subsequently transformed into a subapical cortical ring as tip growth proceeds (DeMay et al. 2009; Helfer and Gladfelter 2006; Kaufmann and Philippsen 2009).

B. CAR Assembly in the *Pezizomycotina*

The CAR is also a key constituent of the cytokinetic machinery required for septum formation in the ***Pezizomycotina*** fungi (Seiler and Justa-Schuch 2010; Berepiki et al. 2011). The dynamics of the actin cytoskeleton during septum formation has recently been addressed using Lifeact (a marker for labeling f-actin in living cells; Riedl et al. 2008) in *N. crassa* (Delgado-Alvarez et al. 2010, 2014; Berepiki et al. 2010). A prominent tangle of actin filaments, the septal actomyosin tangle (SAT), occurs 3–4 min prior to the formin BNI-1 and the anillin BUD-4 at sites of future septum formation. Thus, formin- and anillin-dependent f-actin nucleation and organization seems to be of minor importance for SAT formation, while the appearance of BUD-4 and BNI-1 coincides with maturation of the CAR from the SAT (Fig. 4b). Another striking observation is that the SAT may primarily be generated by transferring existing filaments from established subapical septa to the future site of septation. It is currently unclear if this transfer is based on treadmilling of filaments or on myosin-dependent transport of f-actin.

Moreover, a cortical double ring of patches consisting of components of the Arp2/3 complex, fimbrin and coronin, appear after the coalescence of Lifeact-GFP-labeled cables into a sharp ring, a few seconds before the membrane contraction begins and disappear after CAR contraction has terminated (Delgado-Alvarez et al. 2010, 2014; Echauri-Espinosa et al. 2012; Upadhyay and Shaw 2008). These patches are likely part of the endocytosis machinery that may contribute to membrane remodeling and recycling of supernumerary cell wall enzymes.

Several models are currently discussed that describe the transition from cortical Mid1-containing nodes into the mature CAR in fission yeast (Bathe and Chang 2010; Pollard and Wu 2010; Laporte et al. 2011), a process that seems to correspond to the SAT-CAR transition described for *N. crassa*. These models are non-exclusive and primarily differ in the number and distribution of f-actin nucleating formin assemblies and how the de novo generated filaments coalesce into a stable ring. A recent study also reported that pre-existing f-actin filaments can be recruited in a myosin-dependent manner to cortical actin nodes (Huang et al. 2012). This is consistent with the results obtained in *N. crassa* and analogous findings in animal cells (Chen et al. 2008b; Zhou and Wang 2008), supporting the hypothesis that de novo formin-dependent f-actin assembly at the division site and transport of pre-assembled filaments can contribute to CAR assembly in all eukaryotes. However, the relative contribution of the two actin populations may vary in the different systems.

C. Crosstalk Between the SIN and the MOR Pathways for CAR Regulation

Another fundamental mechanism by which the fission yeast SIN promotes CAR maturation is the inhibition of a competing polarity pathway called the **MORphogenesis network**, which is required for actin organization at cell ends during polarized growth (Gupta and McCollum 2011; Ray et al. 2010). The MOR (called RAM in budding yeast) represents a second Dbf2-related kinase network with an organization similar to that of the SIN (Nelson et al. 2003; Kanai et al. 2005; Das et al. 2009). To promote tip growth, actin is confined to the cell ends where it is required for cell wall deposition. As cells enter mitosis, actin relocates to site of cell division to form the CAR. Since, both processes involve restructuring of the actin cytoskeleton, coordination is presumably important to keep competing actin polarity programs from interfering with each other. Thus, mutual antagonistic function of the SIN and MORphogenesis network is required to coordinate cell growth and division. Inhibition of the MOR is achieved by the SIN effector Sid2, which phosphorylates the MOR kinase Nak1 and the Nak1-associated protein Sog2 in order to block interaction of Nak1-Sog2 with the scaffold protein Mor2 (Gupta et al. 2013). Moreover, the upstream SIN components Cdc7 und Sid1 control and enhance MOR activity in the subsequent interphase after cell division by an unknown mechanism (Kanai et al. 2005).

Tip growth and septum formation occurs simultaneously in filamentous fungi, which requires differential regulation between both NDR kinase pathways (Fig. 4). The localization of *N. crassa* SIN components at the time of SAT to CAR transition suggests the involvement of the SIN in the maturation of a functional CAR (Heilig et al. 2014). A key function of the MOR network in molds seems to inhibit septation initiation (Maerz and Seiler 2010). *N. crassa* MOR mutants produce hyperseptated hyphae, while hyperactivation of the MOR resulted in a reduced septation index (Yarden et al. 1992; Ziv et al. 2009; Seiler and Plamann 2003; Seiler et al. 2006; Maerz et al. 2009). Because the *N. crassa* MOR components do not associate with SPBs (Maerz et al. 2012; Dettmann et al. 2012; Heilig et al. 2014), inhibition of septation initiation by the MOR likely occurs in the cytosol. A second, presumably late function of the MOR during septation might be indicated by the recruitment of all MOR components to the forming septum after initiation of CAR constricting has started (i.e., 2–3 min later than cortex recruitment of the SIN kinase DBF-2; Heilig et al. 2014, Maerz et al. 2012). Such a late function of the MOR is also supported by the analysis of conditional mutants in the MOR effector kinase COT-1, which produce thickened septa and altered septum when grown at restrictive conditions

(Gorovits et al. 2000). These data suggest multiple functions of the MOR in regulating septation initiation and biogenesis of the septum wall in molds.

In line with the opposite functions of the SIN and the MOR during septation, multilevel cross talk between both networks has recently been determined in *N. crassa* (Heilig et al. 2014). The SIN kinase MST-1 regulates the MOR and the SIN both in an enzyme-dependent and -independent manner. MST-1 phosphorylates and activates both the MOR and SIN effector kinases, COT-1 and DBF-2, respectively, by hydrophobic motif phosphorylation. In addition, heterodimerization of the two germinal center kinase MST-1 and the MOR kinase POD-6 inactivates both kinases. An analogous mutually inhibitory interaction occurs between the two NDR kinases DBF-2 and COT-1, which can also form heterodimers. A putative mechanism for kinase inactivation is based on the observation that NDR kinase heterodimerization requires the same protein regions that are required for interaction with their activating MOB protein subunits. Moreover, the kinase-kinase and kinase-MOB interactions were found to be mutually exclusive, and kinase heterodimerization therefore resulted in the displacement of the regulatory subunit and consequently kinase inactivation (Heilig and Seiler; unpublished data).

V. Cell Wall Biogenesis and Cell Division

A. Cell Wall Biogenesis

CAR constriction is coupled with membrane invagination and secretion of membrane-bound cell wall biosynthetic enzymes that build the extracellular septum. Simultaneously, endocytic recycling of supernumerous components (enzymes, membrane, etc.) accompanies septum development. The major enzymes involved in cell wall formation are **chitin and glucan synthases** (Lesage and Bussey 2006; Free 2013). **Chitin** (linear chains of β-1,4-linked *N*-acetylglucosamine molecules) is synthesized by a multifamily group of enzymes that can be grouped into seven classes of chitin synthases (ChS; Riquelme and Bartnicki-García 2008). Most *Pezizomycotina* fungi have single genes for each of the seven reported classes of ChS, while yeasts have a more restricted set of ChS genes. Class II ChS functions as the major enzyme for synthesizing the primary septum in budding yeast (Chs2p; Schmidt et al. 2002) and in *C. albicans* (Chs1; Munro et al. 2001; Walker et al. 2013), while the other ChS in the *Saccharomycotina* are involved in the biosynthesis of the general cell wall and in repair functions (Lesage and Bussey 2006). Although the three-layered septum structure is also described for fission yeast, its primary septum is mainly composed of **β-1,3-glucan** (Humbel et al. 2001; Sugawara et al. 2003). The secondary septum in both yeast models and in *C. albicans* is built by several layers of β-1,3-glucans and mannoproteins (Lesage and Bussey 2006; Humbel et al. 2001; Sugawara et al. 2003).

Despite considerable effort in the cell wall analysis in the *Pezizomycotina* fungi (Free 2013; Latge and Beauvais 2014), the composition and biosynthesis of their septal walls is still poorly understood. Chitin has been shown to be the major component of the *N. crassa* septum (Hunsley and Gooday 1974). The specific localization of CHS-2 (class II) at the constricting rim of the developing septum (in addition to its localization at the tip growth) in *N. crassa* is consistent with a general function of class II ChS in primary septum formation (Fajardo-Somera et al. 2015). The remaining six *N. crassa* ChS localize along the entire septal plate, suggesting a function of these enzymes in remodeling the primary septum and/or synthesizing additional layers that form the secondary septum (Fajardo-Somera et al. 2015).

Regulation of primary septum formation is best understood in budding yeast. The MEN promotes cytokinesis by influencing multiple pathways involved in CAR constriction and septum formation. For instance, the MEN is involved in targeting the chitin synthase **Chs2p** to the bud neck (Meitinger et al. 2010) and also directly regulates the late cytokinetic components Hof1p and Inn1p, two PCH proteins which are homologs of the *S. pombe* Cdc15 (Sanchez-Diaz et al. 2008; Nishihama et al.

2009; Meitinger et al. 2010, 2011). Chs2p is held in an inactive state in the endoplasmatic reticulum by phosphorylation through Cdc28p/CDK1 that blocks its interaction with the COPII component Sec24p and sorting into COPII vesicles (Chuang and Schekman 1996; Zhang et al. 2006; Meitinger et al. 2010; Teh et al. 2009; Jakobsen et al. 2013), while MEN-controlled dephosphorylation by Cdc14p allows Chs2p to enter the secretory pathway (Chin et al. 2012). Subsequent inactivation of Chs2p is achieved through phosphorylation by the MEN effector kinase Dbf2p (Oh et al. 2012).

The second major component of the fungal cell wall is β-1,3-glucan. Its synthesis is catalyzed by membrane-bound β-1,3-glucan synthase complexes, which consist of the catalytic FKS and regulatory Rho1 GTPase subunit (Mazur and Baginsky 1996; Beauvais et al. 2001). Both yeasts have multiple, partially redundant **FKS genes**, and glucan synthesis is essential for viability. This is also the case for the dimorphic *Saccharomycotina* species *C. albicans* (Munro 2013). Most *Pezizomycotina* fungi have only a single FKS gene. Although all components of the β-1,3-glucan synthase complex as well as putative regulators localize to sites of polar growth at hyphal tips, emerging branches and along septa in *N. crassa* and other molds (Vogt and Seiler 2008; Verdin et al. 2009; Richthammer et al. 2012; Sanchez-Leon and Riquelme 2015), β-1,3-glucan synthesis is not essential in *A. fumigatus* (Dichtl et al. 2015), suggesting compensatory functions by other cell wall components in the *Pezizomycotina* fungi.

B. Septum Formation and Cell Division During Fungal Development

Premature activation of the fission yeast MOR (e.g., by misregulation of the SIN) results in inappropriate septum degradation and consequently cell lysis (Gupta et al. 2014). Degradation of the primary septum to trigger cell separation is initiated by RAM-dependent activation of the **transcription factor Ace2p** that controls the expression of Cts1p and Eng1p, the major cell wall-degrading chitinase and glucanase, respectively (Dohrmann et al. 1992; Colman-Lerner et al. 2001; Weiss et al. 2002). *C. albicans* Ace2 mutants also display cell separation defects (Kelly et al. 2004), and, consequently, transcription factor function is inhibited by the filament-inducing transcription factor Efg1 in order to inhibit cell separation after septum formation in hyphae (Wang et al. 2009; Saputo et al. 2014). Budding yeast Ace2p has a paralog, Swi5p, a transcription factor that is primarily involved in cell cycle regulation. Therefore it is currently unclear, if the *Pezizomycotina* fungi have a functional homolog of budding yeast Ace2p. In addition, we have no data that indicate transcriptional regulation by the MOR during tip growth in these fungi, and thus it remains open, if transcriptional regulation impacts vegetative septum formation in the *Pezizomycotina* clade.

Asexual sporulation (conidium) in the *Pezizomycotina* fungi involves the formation of conidia, formed on specialized structures called conidiophores. This is, in principle, achieved by two distinct sporulation patterns exemplified by *A. nidulans* and *N. crassa*, respectively (Park and Yu 2012). During basipetal sporulation the spore forms at the base of a chain and pushes the older cells of the conidial chain away from the spore-forming region, while the acropetal sporulation pattern refers to the fact that the most recently formed spore is at the tip of a chain of spores. Thus, the machinery required for septum formation and cell division must be targeted accordingly. Conidiation in molds is analogous to cell separation in the unicellular yeasts and requires full cell separation through a multilayered cell wall, which is followed by the digestion of the primary cell wall material between two completely formed secondary septum to release mature spores (Springer and Yanofsky 1989; Adams et al. 1998). Consequently, all *A. nidulans* and *N. crassa* mutants of currently characterized proteins required for septum formation in vegetative hyphae are also aconidiate (Rasmussen and Glass 2005; Maerz et al. 2009; Justa-Schuch et al. 2010; Dvash et al. 2010; Heilig et al. 2013; Bruno et al. 2001; Kim et al. 2006, 2009; Si et al. 2010). We have very limited mechanistic insights how the cell division machinery is reprogrammed during the two basic developmental cell division patterns observed in molds. It has recently been shown that the *A. nidulans* homolog of the budding yeast axial bud-site landmark component Axl2p has no obvious role during vegetative growth (Si et al. 2012). **Axl2** is specifically required for the regulation of phialide morphogenesis during conidium, where it appears to promote the recruitment of septins, and Axl2 mutants fail to produce the long chains of conidium. Consistently, Axl2 specifically localized to the phialide-spore junction, implying Axl2 as landmark for reorientation of the division pattern from acropetal growth during phialide formation to basipetal growth during sporulation (Si et al. 2012).

All fungal and animal anillin (with the exception of the *S. pombe* landmark Mid1) interact with the septin scaffold. However, the timing of this interaction and its importance for cell division varies between the different fungal groups (Fig. 5). In vegetative *A. nidulans*, *A. fumigatus*, and *N. crassa* hyphae, the septin play only a minor role during septum formation, and the three fungi form proper septa when any of the five septin genes is deleted (Lindsey et al. 2010; Hernandez-Rodriguez et al. 2012; Berepiki and Read 2013; Vargas-Muniz et al. 2015). In contrast, septin deletion strains of both *Aspergilli* result in major defects in conidiophore development. Similarly, *N. crassa* core septin deletion mutants produce chains of unseparated conidium (Berepiki and Read 2013). In summary, these data are consistent with a late-acting function of the septins during cell separation, which—in molds—is suppressed during vegetative growth.

Intriguingly, splice variants of *C. albicans* transcription factor have recently been associated with RAM/MOR-dependent coordination of septin dynamics and regulation of the incorporation of the Sep7 septin into hyphal septal rings in order to avoid cell separation (Calderon-Norena et al. 2015). Thus, septin dynamics during conidial development in molds may also be regulated through NDR kinase signals. Mutants, in which the activity of the MOR effector kinase COT-1 is reduced, result in conidial separation defect (Ziv et al. 2009), indicating that the MOR functions during developmental cell separation in molds as described for vegetative cell separation in the two yeasts. Interestingly, an *A. fumigatus* Ace2 mutant displayed pleiotropic defects that were primarily associated with conidium morphogenesis and resulted in reduced amounts of generated conidial with highly thickened conidial walls (Ejzykowicz et al. 2009). Thus, MOR-dependent transcriptional regulation might be inhibited during vegetative growth to inhibit cell separation but is induced during the developmental program.

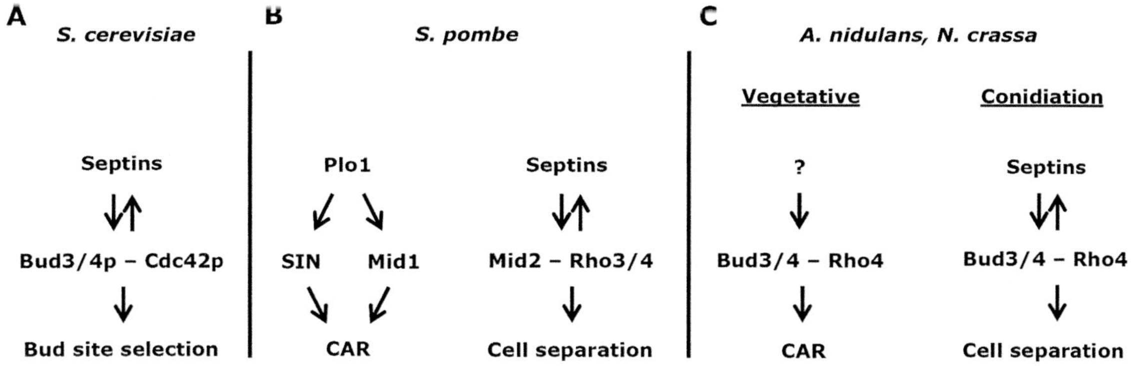

Fig. 5 Putative relationships between anillin-Rho GTPase complexes and the septins. (**a**) The interplay between the septins and the Bud3/4p-Cdc42p complex regulates bud-site selection in budding yeast. (**b**) A similar Mid2-Rho3/4 complex functions at a late stage of cell division to trigger cell separation in fission yeast. In addition, Mid1 has a unique function during septum placement in *S. pombe*. (**c**) The Bud3/4-Rho4 complex is important for CAR formation during vegetative growth of *A. nidulans* and *N. crassa* hyphae. The septins play only a minor role at this stage of the life cycle yet become critical for cell separation during asexual development

Sexual development is also affected in septum mutants. For example, *N. crassa* mutants are female sterile and do not form protoperithecia (Rasmussen and Glass 2005; Maerz et al. 2009; Justa-Schuch et al. 2010; Heilig et al. 2013). Moreover, homozygous crosses of septum strains, in which the female partner has been sheltered by a helper strain, are barren and produce very few ascospores. Interestingly, no septa are formed in ascogenous hyphae in these mutants, indicating multiple developmental defects of theses strains (Rasmussen and Glass 2005; Maerz et al. 2009). In budding yeast and fission yeast, the SIN has recently been shown to be dispensable for progression through meiosis but required for subsequent spore wall formation and ascospore morphology (Krapp et al. 2006; Attner and Amon 2012). Similarly, *N. crassa* SIN mutants produce few but giant ascospores containing all eight nuclei derived from the two meiotic and one mitotic divisions (Raju and Newmeyer 1977; Freitag et al. 2004; Maerz et al. 2009; Heilig et al. 2013). However, these nuclei are then enclosed in a single giant ascospore, supporting an essential function of the SIN during cross wall formation in vegetative cells and during the formation of asco- as well as conidiospores in unicellular and filamentous ascomycetes.

C. Septal Pore-Associated Functions in Filamentous Ascomycete Fungi

The septa of most *Pezizomycotina* species are perforated by simple pores of 350–500 nm in diameter, which allow nuclei, organelles, and cytoplasm to move between compartments (Hunsley and Gooday 1974; Mourino-Perez and Riquelme 2013). However, the mechanisms that terminate CAR constriction and incomplete cell separation are virtually unexplored. The development of certain animal tissues also requires incomplete cytokinesis and the formation of syncytia—a process that is very poorly understood. For instance, in many species, including humans, germ cells remain connected by intercellular bridges, which are required for germ cell development, and their absence results in infertility (Haglund et al. 2011). The robust genetic and cell biological tractability of filamentous fungi thus provides an unparalleled opportunity to determine mechanisms that control complete versus incomplete cytokinesis and the regulated gating of intercellular bridges/septum pores.

The presence of septa is essential for maintaining colony integrity after hyphal injury by rapid plugging of septal pores through peroxisome-derived Woronin bodies (Jedd and Chua 2000; Pieuchot and Jedd 2012). Moreover, the structure and composition of the septum and the connectivity status of septum in filamentous ascomycotina fungi varies within the developing mycelium. This allows age-dependent plugging of interior regions of the colony (e.g., in *N. crassa*: Trinci and Collinge 1973; Hunsley and Gooday 1974) and can establish cellular heterogeneity (e.g., in *A. niger*: de Bekker et al. 2011; Vinck et al. 2011; Wosten et al. 2013). Moreover, dynamic pore closure can be important to separate mitotically active versus inactive compartments (e.g., in *A. nidulans*: Shen et al. 2014) or to compartmentalize communicating regions within the colony (e.g., in *N. crassa*: Dettmann et al. 2014; Jonkers et al. 2014). How pore closure is regulated and which components are involved is poorly understood. Key candidate proteins that are likely involved in this process in varying composition are HEX-1-derived Woronin bodies (Beck et al. 2013), the SOFT protein, which is also involved in cell-cell communication and has recently been identified as scaffold of the cell wall integrity MAP kinase pathway (Fleissner and Glass 2007; Teichert et al. 2014) and a set of septum-associated intrinsically disordered proteins (Lai et al. 2012; Shen et al. 2014). How trafficking of growth-associated factors such as small secretory vesicles toward the tip growth is affected by pore closure is unknown. It also remains possible that more selective transport is still allowed although the pore is closed for bulk transport. In this context, it is also worth noting that a microtubule-organizing center associates with the septum pore in *A. nidulans* (Veith et al. 2005; Zekert et al. 2010; Takeshita and Fischer 2011), which may promote such intercompartmental transport in addition to its implication as component of a speculative cell size-sensing mechanism discussed in Sect. II-B.

VI. Conclusions and Perspectives

Although septum formation is essential for cell proliferation and fungal development, many important questions remain to be addressed. Of particular importance is the identification of signals that determine septum placement in a syncytial compartment. Do specific landmark proteins exist? Are nuclear and cell end-dependent signals involved? What is the function of sMTOCs in this context, and do MTs have any role in sensing of the apical cell size and site selection for septum placement? A second battery of questions involve the regulation of the actin cytoskeleton: how is SAT formation regulated, and what is the relative importance of de novo f-actin nucleation versus assembly of preformed filaments originating from previously formed septum? What is the trigger for CAR constriction and how is CAR constriction terminated? Third, how is polar tip growth and subapical septum formation regulated, if both processes coexist in the filament but depend on the same growth machinery? Forth, why do so many (unrelated) signaling modules associate with the mature septal pore, and how is cell-cell connectivity regulated? Finally, we need to understand how these basic processes are remodeled during fungal development and multicellular differentiation. The answers to these questions will require comparative approaches, and the acknowledgement that yeast and filamentous lifestyles have reused conserved molecular modules in different contexts. In the long term, this will not only improve our understanding of septum formation in vegetative hyphae but also of cell differentiation during ascomycete development and will shed light on the evolutionary consequences of cell compartmentalization when compared to other fungal phyla such as the aseptate zygomycetes or the basidiomycetes that generate highly complex septa.

Acknowledgment Research of the Seiler laboratory is supported through the Deutsche Forschungsgemeinschaft DFG (grants SE1054/4-2, SE1054/6-2, SE1054/7-2, SE1054/9-1).

References

Adams TH, Wieser JK, Yu JH (1998) Asexual sporulation in *Aspergillus nidulans*. Microbiol Mol Biol Rev 62:35–54

Alberti-Segui C, Dietrich F, Altmann-Johl R, Hoepfner D, Philippsen P (2001) Cytoplasmic dynein is required to oppose the force that moves nuclei towards the hyphal tip in the filamentous ascomycete *Ashbya gossypii*. J Cell Sci 114:975–986

Almonacid M, Moseley JB, Janvore J, Mayeux A, Fraisier V, Nurse P, Paoletti A (2009) Spatial control of cytokinesis by Cdr2 kinase and Mid1/anillin nuclear export. Curr Biol 19:961–966

Attner MA, Amon A (2012) Control of the mitotic exit network during meiosis. Mol Biol Cell 23:3122–3132

Bachewich C, Masker K, Osmani S (2005) The polo-like kinase PLKA is required for initiation and progression through mitosis in the filamentous fungus *Aspergillus nidulans*. Mol Microbiol 55:572–587

Bähler J, Steever AB, Wheatley S, Wang Y, Pringle JR, Gould KL, McCollum D (1998) Role of polo kinase and Mid1p in determining the site of cell division in fission yeast. J Cell Biol 143:1603–1616

Balasubramanian MK, Bi E, Glotzer M (2004) Comparative analysis of cytokinesis in budding yeast, fission yeast and animal cells. Curr Biol 14:R806–R818

Barr FA, Gruneberg U (2007) Cytokinesis: placing and making the final cut. Cell 131:847–860

Bathe M, Chang F (2010) Cytokinesis and the contractile ring in fission yeast: towards a systems-level understanding. Trends Microbiol 18:38–45

Beauvais A, Bruneau JM, Mol PC, Buitrago MJ, Legrand R, Latge JP (2001) Glucan synthase complex of *Aspergillus fumigatus*. J Bacteriol 183:2273–2279

Beck J, Echtenacher B, Ebel F (2013) Woronin bodies, their impact on stress resistance and virulence of the pathogenic mould *Aspergillus fumigatus* and their anchoring at the septal pore of filamentous Ascomycota. Mol Microbiol 89:857–871

Berepiki A, Read ND (2013) Septins are important for cell polarity, septation and asexual spore formation in *Neurospora crassa* and show different patterns of localisation at germ tube tips. PLoS One 8: e63843

Berepiki A, Lichius A, Shoji JY, Tilsner J, Read ND (2010) F-actin dynamics in *Neurospora crassa*. Eukaryot Cell 9:547–557

Berepiki A, Lichius A, Read ND (2011) Actin organization and dynamics in filamentous fungi. Nat Rev Microbiol 9:876–887

Berlin A, Paoletti A, Chang F (2003) Mid2p stabilizes septin rings during cytokinesis in fission yeast. J Cell Biol 160:1083–1092

Bi E, Park HO (2012) Cell polarization and cytokinesis in budding yeast. Genetics 191:347–387
Bi E, Maddox P, Lew DJ, Salmon ED, McMillan JN, Yeh E, Pringle JR (1998) Involvement of an actomyosin contractile ring in *Saccharomyces cerevisiae* cytokinesis. J Cell Biol 142:1301–1312
Blackwell M (2011) The fungi: 1, 2, 3 ... 5.1 million species? Am J Bot 98:426–438
Bloemendal S, Lord KM, Rech C, Hoff B, Engh I, Read ND, Kuck U (2010) A mutant defective in sexual development produces aseptate ascogonia. Eukaryot Cell 9:1856–1866
Bloemendal S, Bernhards Y, Bartho K, Dettmann A, Voigt O, Teichert I, Seiler S, Wolters DA, Poggeler S, Kuck U (2012) A homologue of the human STRIPAK complex controls sexual development in fungi. Mol Microbiol 84:310–323
Bohnert KA, Willet AH, Kovar DR, Gould KL (2013) Formin-based control of the actin cytoskeleton during cytokinesis. Biochem Soc Trans 41:1750–1754
Bruno KS, Morrell JL, Hamer JE, Staiger CJ (2001) SEPH, a Cdc7p orthologue from *Aspergillus nidulans*, functions upstream of actin ring formation during cytokinesis. Mol Microbiol 42:3–12
Cabernard C, Prehoda KE, Doe CQ (2010) A spindle-independent cleavage furrow positioning pathway. Nature 467:91–94
Calderon-Norena DM, Gonzalez-Novo A, Orellana-Munoz S, Gutierrez-Escribano P, Arnaiz-Pita Y, Duenas-Santero E, Suarez MB, Bougnoux ME, Del Rey F, Sherlock G, d'Enfert C, Correa-Bordes J, de Aldana CR (2015) A single nucleotide polymorphism uncovers a novel function for the transcription factor Ace2 during *Candida albicans* hyphal development. PLoS Genet 11:e1005152
Celton-Morizur S, Racine V, Sibarita JB, Paoletti A (2006) Pom1 kinase links division plane position to cell polarity by regulating Mid1p cortical distribution. J Cell Sci 119:4710–4718
Chang F, Peter M (2003) Yeasts make their mark. Nat Cell Biol 5:294–299
Chant J, Herskowitz I (1991) Genetic control of bud site selection in yeast by a set of gene products that constitute a morphogenetic pathway. Cell 65:1203–1212
Chen CT, Feoktistova A, Chen JS, Shim YS, Clifford DM, Gould KL, McCollum D (2008a) The SIN kinase Sid2 regulates cytoplasmic retention of the S. pombe Cdc14-like phosphatase Clp1. Curr Biol 18:1594–1599
Chen W, Foss M, Tseng KF, Zhang D (2008b) Redundant mechanisms recruit actin into the contractile ring in silkworm spermatocytes. *PLoS Biol* 6:e209
Chin CF, Bennett AM, Ma WK, Hall MC, Yeong FM (2012) Dependence of Chs2 ER export on dephosphorylation by cytoplasmic Cdc14 ensures that septum formation follows mitosis. Mol Biol Cell 23:45–58
Chuang JS, Schekman RW (1996) Differential trafficking and timed localization of two chitin synthase proteins, Chs2p and Chs3p. J Cell Biol 135:597–610
Clifford DM, Chen CT, Roberts RH, Feoktistova A, Wolfe BA, Chen JS, McCollum D, Gould KL (2008) The role of Cdc14 phosphatases in the control of cell division. Biochem Soc Trans 36:436–438
Colman-Lerner A, Chin TE, Brent R (2001) Yeast Cbk1 and Mob2 activate daughter-specific genetic programs to induce asymmetric cell fates. Cell 107:739–750
Daga RR, Chang F (2005) Dynamic positioning of the fission yeast cell division plane. Proc Natl Acad Sci U S A 102:8228–8232
Das M, Wiley DJ, Chen X, Shah K, Verde F (2009) The conserved NDR kinase Orb6 controls polarized cell growth by spatial regulation of the small GTPase Cdc42. Curr Biol 19:1314–1319
D'Avino PP (2009) How to scaffold the contractile ring for a safe cytokinesis – lessons from anillin-related proteins. J Cell Sci 122:1071–1079
D'Avino PP, Takeda T, Capalbo L, Zhang W, Lilley KS, Laue ED, Glover DM (2008) Interaction between anillin and RacGAP50C connects the actomyosin contractile ring with spindle microtubules at the cell division site. J Cell Sci 121:1151–1158
de Bekker C, Bruning O, Jonker MJ, Breit TM, Wosten HA (2011) Single cell transcriptomics of neighboring hyphae of *Aspergillus niger*. Genome Biol 12:R71
De Souza CP, Osmani AH, Wu LP, Spotts JL, Osmani SA (2000) Mitotic histone H3 phosphorylation by the NIMA kinase in *Aspergillus nidulans*. Cell 102:293–302
De Souza CP, Osmani AH, Hashmi SB, Osmani SA (2004) Partial nuclear pore complex disassembly during closed mitosis in *Aspergillus nidulans*. Curr Biol 14:1973–1984
De Souza CP, Hashmi SB, Osmani AH, Osmani SA (2014) Application of a new dual localization-affinity purification tag reveals novel aspects of protein kinase biology in *Aspergillus nidulans*. PLoS One 9:e90911
Delgado-Alvarez DL, Callejas-Negrete OA, Gomez N, Freitag M, Roberson RW, Smith LG, Mourino-Perez RR (2010) Visualization of F-actin localization and dynamics with live cell markers in *Neurospora crassa*. Fungal Genet Biol 47:573–586
Delgado-Alvarez DL, Bartnicki-Garcia S, Seiler S, Mourino-Perez RR (2014) Septum development in *Neurospora crassa*: the septal actomyosin tangle. PLoS One 9:e96744
DeMay BS, Meseroll RA, Occhipinti P, Gladfelter AS (2009) Regulation of distinct septin rings in a single cell by Elm1p and Gin4p kinases. Mol Biol Cell 20:2311–2326
Dettmann A, Illgen J, Marz S, Schurg T, Fleissner A, Seiler S (2012) The NDR kinase scaffold HYM1/MO25 is essential for MAK2 map kinase signaling in *Neurospora crassa*. PLoS Genet 8:e1002950

Dettmann A, Heilig Y, Ludwig S, Schmitt K, Illgen J, Fleissner A, Valerius O, Seiler S (2013) HAM-2 and HAM-3 are central for the assembly of the Neurospora STRIPAK complex at the nuclear envelope and regulate nuclear accumulation of the MAP kinase MAK-1 in a MAK-2-dependent manner. Mol Microbiol 90:796–812

Dettmann A, Heilig Y, Valerius O, Ludwig S, Seiler S (2014) Fungal communication requires the MAK-2 pathway elements STE-20 and RAS-2, the NRC-1 adapter STE-50 and the MAP kinase scaffold HAM-5. PLoS Genet 10:e1004762

Dichtl K, Samantaray S, Aimanianda V, Zhu Z, Prevost MC, Latge JP, Ebel F, Wagener J (2015) *Aspergillus fumigatus* devoid of cell wall beta-1,3-glucan is viable, massively sheds galactomannan and is killed by septum formation inhibitors. Mol Microbiol 95:458–471

Dietrich FS, Voegeli S, Brachat S, Lerch A, Gates K, Steiner S, Mohr C, Pohlmann R, Luedi P, Choi S, Wing RA, Flavier A, Gaffney TD, Philippsen P (2004) The *Ashbya gossypii* genome as a tool for mapping the ancient *Saccharomyces cerevisiae* genome. Science 304:304–307

Dohrmann PR, Butler G, Tamai K, Dorland S, Greene JR, Thiele DJ, Stillman DJ (1992) Parallel pathways of gene regulation: homologous regulators SWI5 and ACE2 differentially control transcription of HO and chitinase. Genes Dev 6:93–104

Dvash E, Kra-Oz G, Ziv C, Carmeli S, Yarden O (2010) The NDR kinase DBF-2 is involved in regulation of mitosis, conidial development, and glycogen metabolism in *Neurospora crassa*. Eukaryot Cell 9:502–513

Echauri-Espinosa RO, Callejas-Negrete OA, Roberson RW, Bartnicki-Garcia S, Mourino-Perez RR (2012) Coronin is a component of the endocytic collar of hyphae of *Neurospora crassa* and is necessary for normal growth and morphogenesis. PLoS One 7: e38237

Eggert US, Mitchison TJ, Field CM (2006) Animal cytokinesis: from parts list to mechanisms. Annu Rev Biochem 75:543–566

Ejzykowicz DE, Cunha MM, Rozental S, Solis NV, Gravelat FN, Sheppard DC, Filler SG (2009) The *Aspergillus fumigatus* transcription factor Ace2 governs pigment production, conidiation and virulence. Mol Microbiol 72:155–169

El Amine N, Kechad A, Jananji S, Hickson GR (2013) Opposing actions of septins and Sticky on anillin promote the transition from contractile to midbody ring. J Cell Biol 203:487–504

Eluere R, Varlet I, Bernadac A, Simon MN (2012) Cdk and the anillin homolog Bud4 define a new pathway regulating septin organization in yeast. Cell Cycle 11:151–158

Fajardo-Somera RA, Johnk B, Bayram O, Valerius O, Braus GH, Riquelme M (2015) Dissecting the function of the different chitin synthases in vegetative growth and sexual development in *Neurospora crassa*. Fungal Genet Biol 75:30–45

Fankhauser C, Simanis V (1994) The cdc7 protein kinase is a dosage dependent regulator of septum formation in fission yeast. EMBO J 13:3011–3019

Feoktistova A, Morrell-Falvey J, Chen JS, Singh NS, Balasubramanian MK, Gould KL (2012) The fission yeast septation initiation network (SIN) kinase, Sid2, is required for SIN asymmetry and regulates the SIN scaffold, Cdc11. Mol Biol Cell 23:1636–1645

Field CM, Alberts BM (1995) Anillin, a contractile ring protein that cycles from the nucleus to the cell cortex. J Cell Biol 131:165–178

Finlayson MR, Helfer-Hungerbuhler AK, Philippsen P (2011) Regulation of exit from mitosis in multinucleate *Ashbya gossypii* cells relies on a minimal network of genes. Mol Biol Cell 22:3081–3093

Finley KR, Berman J (2005) Microtubules in *Candida albicans* hyphae drive nuclear dynamics and connect cell cycle progression to morphogenesis. Eukaryot Cell 4:1697–1711

Finley KR, Bouchonville KJ, Quick A, Berman J (2008) Dynein-dependent nuclear dynamics affect morphogenesis in *Candida albicans* by means of the Bub2p spindle checkpoint. J Cell Sci 121:466–476

Fischer R, Zekert N, Takeshita N (2008) Polarized growth in fungi–interplay between the cytoskeleton, positional markers and membrane domains. Mol Microbiol 68:813–826

Fleissner A, Glass NL (2007) SO, a protein involved in hyphal fusion in *Neurospora crassa*, localizes to septal plugs. Eukaryot Cell 6:84–94

Free SJ (2013) Fungal cell wall organization and biosynthesis. Adv Genet 81:33–82

Freitag M, Hickey PC, Raju NB, Selker EU, Read ND (2004) GFP as a tool to analyze the organization, dynamics and function of nuclei and microtubules in *Neurospora crassa*. Fungal Genet Biol 41:897–910

Frey S, Reschka EJ, Poggeler S (2015) Germinal center kinases SmKIN3 and SmKIN24 are associated with the *Sordaria macrospora* striatin-interacting phosphatase and kinase (STRIPAK) complex. PLoS One 10:e0139163

Gale C, Gerami-Nejad M, McClellan M, Vandoninck S, Longtine MS, Berman J (2001) *Candida albicans* Int1p interacts with the septin ring in yeast and hyphal cells. Mol Biol Cell 12:3538–3549

Gao XD, Sperber LM, Kane SA, Tong Z, Tong AH, Boone C, Bi E (2007) Sequential and distinct roles of the cadherin domain-containing protein Axl2p in cell polarization in yeast cell cycle. Mol Biol Cell 18:2542–2560

Gladfelter AS, Pringle JR, Lew DJ (2001) The septin cortex at the yeast mother-bud neck. Curr Opin Microbiol 4:681–689

Gonzalez-Novo A, Correa-Bordes J, Labrador L, Sanchez M, Vazquez de Aldana CR, Jimenez J (2008) Sep7 is essential to modify septin ring dynamics

and inhibit cell separation during *Candida albicans* hyphal growth. Mol Biol Cell 19:1509–1518
Gonzalez-Novo A, Labrador L, Pablo-Hernando ME, Correa-Bordes J, Sanchez M, Jimenez J, Vazquez de Aldana CR (2009) Dbf2 is essential for cytokinesis and correct mitotic spindle formation in *Candida albicans*. Mol Microbiol 72:1364–1378
Gorovits R, Sjollema KA, Sietsma JH, Yarden O (2000) Cellular distribution of COT1 kinase in *Neurospora crassa*. Fungal Genet Biol 30:63–70
Govindaraghavan M, McGuire Anglin SL, Shen KF, Shukla N, De Souza CP, Osmani SA (2014) Identification of interphase functions for the NIMA kinase involving microtubules and the ESCRT pathway. PLoS Genet 10:e1004248
Grallert A, Connolly Y, Smith DL, Simanis V, Hagan IM (2012) The *S. pombe* cytokinesis NDR kinase Sid2 activates Fin1 NIMA kinase to control mitotic commitment through Pom1/Wee1. Nat Cell Biol 14:738–745
Gregory SL, Ebrahimi S, Milverton J, Jones WM, Bejsovec A, Saint R (2008) Cell division requires a direct link between microtubule-bound RacGAP and anillin in the contractile ring. Curr Biol 18:25–29
Gu Y, Oliferenko S (2015) Comparative biology of cell division in the fission yeast clade. Curr Opin Microbiol 28:18–25
Gu Y, Yam C, Oliferenko S (2015) Rewiring of cellular division site selection in evolution of fission yeasts. Curr Biol 25:1187–1194
Gull K (1978) Form and function of septa in filamentous fungi. In: Smith JE, Berry DR (eds) The filamentous fungi, Developmental mycology. Wiley, New York, pp 78–93
Gupta S, McCollum D (2011) Crosstalk between NDR kinase pathways coordinates cell cycle dependent actin rearrangements. Cell Div 6:19
Gupta S, Mana-Capelli S, McLean JR, Chen CT, Ray S, Gould KL, McCollum D (2013) Identification of SIN pathway targets reveals mechanisms of crosstalk between NDR kinase pathways. Curr Biol 23:333–338
Gupta S, Govindaraghavan M, McCollum D (2014) Cross talk between NDR kinase pathways coordinates cytokinesis with cell separation in *Schizosaccharomyces pombe*. Eukaryot Cell 13:1104–1112
Hachet O, Simanis V (2008) Mid1p/anillin and the septation initiation network orchestrate contractile ring assembly for cytokinesis. Genes Dev 22:3205–3216
Haglund K, Nezis IP, Stenmark H (2011) Structure and functions of stable intercellular bridges formed by incomplete cytokinesis during development. Commun Integr Biol 4:1–9
Harris SD (2001) Septum formation in *Aspergillus nidulans*. Curr Opin Microbiol 4:736–739
Harris SD, Kraus PR (1998) Regulation of septum formation in *Aspergillus nidulans* by a DNA damage checkpoint pathway. Genetics 148:1055–1067
Harris SD, Morrell JL, Hamer JE (1994) Identification and characterization of *Aspergillus nidulans* mutants defective in cytokinesis. Genetics 136:517–532
Heilig Y, Schmitt K, Seiler S (2013) Phospho-regulation of the *Neurospora crassa* septation initiation network. PLoS One 8:e79464
Heilig Y, Dettmann A, Mourino-Perez RR, Schmitt K, Valerius O, Seiler S (2014) Proper actin ring formation and septum constriction requires coordinated regulation of SIN and MOR pathways through the germinal centre kinase MST-1. PLoS Genet 10:e1004306
Helfer H, Gladfelter AS (2006) AgSwe1p regulates mitosis in response to morphogenesis and nutrients in multinucleated *Ashbya gossypii* cells. Mol Biol Cell 17:4494–4512
Hernandez-Rodriguez Y, Hastings S, Momany M (2012) The septin AspB in *Aspergillus nidulans* forms bars and filaments and plays roles in growth emergence and conidiation. Eukaryot Cell 11:311–323
Hickson GR, O'Farrell PH (2008) Rho-dependent control of anillin behavior during cytokinesis. J Cell Biol 180:285–294
Higashitsuji Y, Herrero S, Takeshita N, Fischer R (2009) The cell end marker protein TeaC is involved in growth directionality and septation in *Aspergillus nidulans*. Eukaryot Cell 8:957–967
Hill TW, Jackson-Hayes L, Wang X, Hoge BL (2015) A mutation in the converter subdomain of *Aspergillus nidulans* MyoB blocks constriction of the actomyosin ring in cytokinesis. Fungal Genet Biol 75:72–83
Hong CI, Zamborszky J, Baek M, Labiscsak L, Ju K, Lee H, Larrondo LF, Goity A, Chong HS, Belden WJ, Csikasz-Nagy A (2014) Circadian rhythms synchronize mitosis in *Neurospora crassa*. Proc Natl Acad Sci U S A 111:1397–1402
Horio T, Oakley BR (2005) The role of microtubules in rapid hyphal tip growth of *Aspergillus nidulans*. Mol Biol Cell 16:918–926
Hou MC, Guertin DA, McCollum D (2004) Initiation of cytokinesis is controlled through multiple modes of regulation of the Sid2p-Mob1p kinase complex. Mol Cell Biol 24:3262–3276
Huang Y, Chew TG, Ge W, Balasubramanian MK (2007) Polarity determinants Tea1p, Tea4p, and Pom1p inhibit division-septum assembly at cell ends in fission yeast. Dev Cell 12:987–996
Huang Y, Yan H, Balasubramanian MK (2008) Assembly of normal actomyosin rings in the absence of Mid1p and cortical nodes in fission yeast. J Cell Biol 183:979–988
Huang J, Huang Y, Yu H, Subramanian D, Padmanabhan A, Thadani R, Tao Y, Tang X, Wedlich-Soldner R, Balasubramanian MK (2012) Nonmedially assembled F-actin cables incorporate into the actomyosin ring in fission yeast. J Cell Biol 199:831–847

Humbel BM, Konomi M, Takagi T, Kamasawa N, Ishijima SA, Osumi M (2001) In situ localization of beta-glucans in the cell wall of *Schizosaccharomyces pombe*. Yeast 18:433–444

Hunsley D, Gooday GW (1974) The structure and development of septa in *Neurospora crassa*. Protoplasma 82:125–146

Jackson SL (2001) Do hyphae pulse as they grow? New Phytol 151:556–560

Jakobsen MK, Cheng Z, Lam SK, Roth-Johnson E, Barfield RM, Schekman R (2013) Phosphorylation of Chs2p regulates interaction with COPII. J Cell Sci 126:2151–2156

James TY, Kauff F, Schoch CL, Matheny PB, Hofstetter V, Cox CJ, Celio G, Gueidan C, Fraker E, Miadlikowska J, Lumbsch HT, Rauhut A, Reeb V, Arnold AE, Amtoft A, Stajich JE, Hosaka K, Sung GH, Johnson D, O'Rourke B, Crockett M, Binder M, Curtis JM, Slot JC, Wang Z, Wilson AW, Schussler A, Longcore JE, O'Donnell K, Mozley-Standridge S, Porter D, Letcher PM, Powell MJ, Taylor JW, White MM, Griffith GW, Davies DR, Humber RA, Morton JB, Sugiyama J, Rossman AY, Rogers JD, Pfister DH, Hewitt D, Hansen K, Hambleton S, Shoemaker RA, Kohlmeyer J, Volkmann-Kohlmeyer B, Spotts RA, Serdani M, Crous PW, Hughes KW, Matsuura K, Langer E, Langer G, Untereiner WA, Lucking R, Budel B, Geiser DM, Aptroot A, Diederich P, Schmitt I, Schultz M, Yahr R, Hibbett DS, Lutzoni F, McLaughlin DJ, Spatafora JW, Vilgalys R (2006) Reconstructing the early evolution of fungi using a six-gene phylogeny. Nature 443:818–822

Jansen JM, Barry MF, Yoo CK, Weiss EL (2006) Phosphoregulation of Cbk1 is critical for RAM network control of transcription and morphogenesis. J Cell Biol 175:755–766

Jedd G, Chua NH (2000) A new self-assembled peroxisomal vesicle required for efficient resealing of the plasma membrane. Nat Cell Biol 2:226–231

Johnson AE, McCollum D, Gould KL (2012) Polar opposites: fine-tuning cytokinesis through SIN asymmetry. Cytoskeleton (Hoboken) 69:686–699

Jonkers W, Leeder AC, Ansong C, Wang Y, Yang F, Starr TL, Camp DG II, Smith RD, Glass NL (2014) HAM-5 functions as a MAP kinase scaffold during cell fusion in *Neurospora crassa*. PLoS Genet 10:e1004783

Justa-Schuch D, Heilig Y, Richthammer C, Seiler S (2010) Septum formation is regulated by the RHO4-specific exchange factors BUD3 and RGF3 and by the landmark protein BUD4 in *Neurospora crassa*. Mol Microbiol 76:220–235

Kaminskyj SG (2000) Septum position is marked at the tip of *Aspergillus nidulans* hyphae. Fungal Genet Biol 31:105–113

Kanai M, Kume K, Miyahara K, Sakai K, Nakamura K, Leonhard K, Wiley DJ, Verde F, Toda T, Hirata D (2005) Fission yeast MO25 protein is localized at SPB and septum and is essential for cell morphogenesis. EMBO J 24:3012–3025

Kang PJ, Angerman E, Jung CH, Park HO (2012) Bud4 mediates the cell-type-specific assembly of the axial landmark in budding yeast. J Cell Sci 125:3840–3849

Kang PJ, Hood-DeGrenier JK, Park HO (2013) Coupling of septins to the axial landmark by Bud4 in budding yeast. J Cell Sci 126:1218–1226

Kang PJ, Lee ME, Park HO (2014) Bud3 activates Cdc42 to establish a proper growth site in budding yeast. J Cell Biol 206:19–28

Kaufmann A, Philippsen P (2009) Of bars and rings: Hof1-dependent cytokinesis in multiseptated hyphae of *Ashbya gossypii*. Mol Cell Biol 29:771–783

Kelly MT, MacCallum DM, Clancy SD, Odds FC, Brown AJ, Butler G (2004) The *Candida albicans* CaACE2 gene affects morphogenesis, adherence and virulence. Mol Microbiol 53:969–983

Kim JM, Lu L, Shao R, Chin J, Liu B (2006) Isolation of mutations that bypass the requirement of the septation initiation network for septum formation and conidiation in *Aspergillus nidulans*. Genetics 173:685–696

Kim JM, Zeng CJ, Nayak T, Shao R, Huang AC, Oakley BR, Liu B (2009) Timely septation requires SNAD-dependent spindle pole body localization of the septation initiation network components in the filamentous fungus *Aspergillus nidulans*. Mol Biol Cell 20:2874–2884

Knechtle P, Dietrich F, Philippsen P (2003) Maximal polar growth potential depends on the polarisome component AgSpa2 in the filamentous fungus *Ashbya gossypii*. Mol Biol Cell 14:4140–4154

Konzack S, Rischitor PE, Enke C, Fischer R (2005) The role of the kinesin motor KipA in microtubule organization and polarized growth of *Aspergillus nidulans*. Mol Biol Cell 16:497–506

Krapp A, Collin P, Cokoja A, Dischinger S, Cano E, Simanis V (2006) The *Schizosaccharomyces pombe* septation initiation network (SIN) is required for spore formation in meiosis. J Cell Sci 119:2882–2891

Kraus PR, Harris SD (2001) The *Aspergillus nidulans* snt genes are required for the regulation of septum formation and cell cycle checkpoints. Genetics 159:557–569

Lai J, Koh CH, Tjota M, Pieuchot L, Raman V, Chandrababu KB, Yang D, Wong L, Jedd G (2012) Intrinsically disordered proteins aggregate at fungal cell-to-cell channels and regulate intercellular connectivity. Proc Natl Acad Sci U S A 109:15781–15786

Laporte D, Zhao R, Wu JQ (2010) Mechanisms of contractile-ring assembly in fission yeast and beyond. Semin Cell Dev Biol 21:892–898

Laporte D, Coffman VC, Lee IJ, Wu JQ (2011) Assembly and architecture of precursor nodes during fission yeast cytokinesis. J Cell Biol 192:1005–1021

Latge JP, Beauvais A (2014) Functional duality of the cell wall. Curr Opin Microbiol 20:111–117
Lee IJ, Wu JQ (2012) Characterization of Mid1 domains for targeting and scaffolding in fission yeast cytokinesis. J Cell Sci 125:2973–2985
Lesage G, Bussey H (2006) Cell wall assembly in *Saccharomyces cerevisiae*. Microbiol Mol Biol Rev 70:317–343
Lindsey R, Cowden S, Hernandez-Rodriguez Y, Momany M (2010) Septins AspA and AspC are important for normal development and limit the emergence of new growth foci in the multicellular fungus *Aspergillus nidulans*. Eukaryot Cell 9:155–163
Lippincott J, Li R (1998) Sequential assembly of myosin II, an IQGAP-like protein, and filamentous actin to a ring structure involved in budding yeast cytokinesis. J Cell Biol 140:355–366
Liu G, Young D (2012) Conserved Orb6 phosphorylation sites are essential for polarized cell growth in *Schizosaccharomyces pombe*. PLoS One 7:e37221
Luo J, Vallen EA, Dravis C, Tcheperegine SE, Drees B, Bi E (2004) Identification and functional analysis of the essential and regulatory light chains of the only type II myosin Myo1p in *Saccharomyces cerevisiae*. J Cell Biol 165:843–855
Lynch EM, Groocock LM, Borek WE, Sawin KE (2014) Activation of the gamma-tubulin complex by the Mto1/2 complex. Curr Biol 24:896–903
Maerz S, Seiler S (2010) Tales of RAM and MOR: NDR kinase signaling in fungal morphogenesis. Curr Opin Microbiol 13:663–671
Maerz S, Dettmann A, Ziv C, Liu Y, Valerius O, Yarden O, Seiler S (2009) Two NDR kinase-MOB complexes function as distinct modules during septum formation and tip extension in *Neurospora crassa*. Mol Microbiol 74:707–723
Maerz S, Dettmann A, Seiler S (2012) Hydrophobic motif phosphorylation coordinates activity and polar localization of the *Neurospora crassa* nuclear Dbf2-related kinase COT1. Mol Cell Biol 32:2083–2098
Mah AS, Jang J, Deshaies RJ (2001) Protein kinase Cdc15 activates the Dbf2-Mob1 kinase complex. Proc Natl Acad Sci U S A 98:7325–7330
Martin SG, Arkowitz RA (2014) Cell polarization in budding and fission yeasts. FEMS Microbiol Rev 38:228–253
Mazur P, Baginsky W (1996) In vitro activity of 1,3-beta-D-glucan synthase requires the GTP-binding protein Rho1. J Biol Chem 271:14604–14609
McLaughlin DJ, Hibbett DS, Lutzoni F, Spatafora JW, Vilgalys R (2009) The search for the fungal tree of life. Trends Microbiol 17:488–497
McMurray MA, Thorner J (2009) Reuse, replace, recycle. Specificity in subunit inheritance and assembly of higher-order septin structures during mitotic and meiotic division in budding yeast. Cell Cycle 8:195–203
Meitinger F, Petrova B, Lombardi IM, Bertazzi DT, Hub B, Zentgraf H, Pereira G (2010) Targeted localization of Inn1, Cyk3 and Chs2 by the mitotic-exit network regulates cytokinesis in budding yeast. J Cell Sci 123:1851–1861
Meitinger F, Boehm ME, Hofmann A, Hub B, Zentgraf H, Lehmann WD, Pereira G (2011) Phosphorylation-dependent regulation of the F-BAR protein Hof1 during cytokinesis. Genes Dev 25:875–888
Meitinger F, Palani S, Pereira G (2012) The power of MEN in cytokinesis. Cell Cycle 11:219–228
Milne SW, Cheetham J, Lloyd D, Shaw S, Moore K, Paszkiewicz KH, Aves SJ, Bates S (2014) Role of *Candida albicans* Tem1 in mitotic exit and cytokinesis. Fungal Genet Biol 69:84–95
Minke PF, Lee IH, Plamann M (1999) Microscopic analysis of Neurospora ropy mutants defective in nuclear distribution. Fungal Genet Biol 28:55–67
Mogilevsky K, Glory A, Bachewich C (2012) The Polo-like kinase PLKA in *Aspergillus nidulans* is not essential but plays important roles during vegetative growth and development. Eukaryot Cell 11:194–205
Momany M, Hamer JE (1997) Relationship of actin, microtubules, and crosswall synthesis during septation in *Aspergillus nidulans*. Cell Motil Cytoskeleton 38:373–384
Morrell JL, Tomlin GC, Rajagopalan S, Venkatram S, Feoktistova AS, Tasto JJ, Mehta S, Jennings JL, Link A, Balasubramanian MK, Gould KL (2004) Sid4p-Cdc11p assembles the septation initiation network and its regulators at the *S. pombe* SPB. Curr Biol 14:579–584
Moseley JB (2015) Cytokinesis: does Mid1 have an identity crisis? Curr Biol 25:R364–R366
Mourino-Perez RR, Riquelme M (2013) Recent advances in septum biogenesis in *Neurospora crassa*. Adv Genet 83:99–134
Munoz S, Manjon E, Sanchez Y (2014) The putative exchange factor Gef3p interacts with Rho3p GTPase and the septin ring during cytokinesis in fission yeast. J Biol Chem 289:21995–22007
Munro CA (2013) Chitin and glucan, the yin and yang of the fungal cell wall, implications for antifungal drug discovery and therapy. Adv Appl Microbiol 83:145–172
Munro CA, Winter K, Buchan A, Henry K, Becker JM, Brown AJ, Bulawa CE, Gow NA (2001) Chs1 of *Candida albicans* is an essential chitin synthase required for synthesis of the septum and for cell integrity. Mol Microbiol 39:1414–1426
Nelson B, Kurischko C, Horecka J, Mody M, Nair P, Pratt L, Zougman A, McBroom LD, Hughes TR, Boone C, Luca FC (2003) RAM: a conserved signaling network that regulates Ace2p transcriptional activity and polarized morphogenesis. Mol Biol Cell 14:3782–3803
Nishihama R, Schreiter JH, Onishi M, Vallen EA, Hanna J, Moravcevic K, Lippincott MF, Han H, Lemmon MA, Pringle JR, Bi E (2009) Role of Inn1 and its

interactions with Hof1 and Cyk3 in promoting cleavage furrow and septum formation in *S. cerevisiae*. J Cell Biol 185:995–1012
Oegema K, Savoian MS, Mitchison TJ, Field CM (2000) Functional analysis of a human homologue of the Drosophila actin binding protein anillin suggests a role in cytokinesis. J Cell Biol 150:539–552
Oh Y, Chang KJ, Orlean P, Wloka C, Deshaies R, Bi E (2012) Mitotic exit kinase Dbf2 directly phosphorylates chitin synthase Chs2 to regulate cytokinesis in budding yeast. Mol Biol Cell 23:2445–2456
Ohkura H, Hagan IM, Glover DM (1995) The conserved *Schizosaccharomyces pombe* kinase plo1, required to form a bipolar spindle, the actin ring, and septum, can drive septum formation in G1 and G2 cells. Genes Dev 9:1059–1073
Padte NN, Martin SG, Howard M, Chang F (2006) The cell-end factor pom1p inhibits mid1p in specification of the cell division plane in fission yeast. Curr Biol 16:2480–2487
Paoletti A, Chang F (2000) Analysis of mid1p, a protein required for placement of the cell division site, reveals a link between the nucleus and the cell surface in fission yeast. Mol Biol Cell 11:2757–2773
Park HS, Yu JH (2012) Multi-copy genetic screen in *Aspergillus nidulans*. Methods Mol Biol 944:183–190
Piekny AJ, Glotzer M (2008) Anillin is a scaffold protein that links RhoA, actin, and myosin during cytokinesis. Curr Biol 18:30–36
Piekny AJ, Maddox AS (2010) The myriad roles of anillin during cytokinesis. Semin Cell Dev Biol 21:881–891
Pieuchot L, Jedd G (2012) Peroxisome assembly and functional diversity in eukaryotic microorganisms. Annu Rev Microbiol 66:237–263
Plamann M, Minke PF, Tinsley JH, Bruno KS (1994) Cytoplasmic dynein and actin-related protein Arp1 are required for normal nuclear distribution in filamentous fungi. J Cell Biol 127:139–149
Pollard TD, Wu JQ (2010) Understanding cytokinesis: lessons from fission yeast. Nat Rev Mol Cell Biol 11:149–155
Pringle A, Taylor J (2002) The fitness of filamentous fungi. Trends Microbiol 10:474–481
Raju NB, Newmeyer D (1977) Giant ascospores and abnormal croziers in a mutant of *Neurospora crassa*. Exp Mycol 1:152–165
Rasmussen CG, Glass NL (2005) A Rho-type GTPase, rho-4, is required for septation in *Neurospora crassa*. Eukaryot Cell 4:1913–1925
Ray S, Kume K, Gupta S, Ge W, Balasubramanian M, Hirata D, McCollum D (2010) The mitosis-to-interphase transition is coordinated by cross talk between the SIN and MOR pathways in *Schizosaccharomyces pombe*. J Cell Biol 190:793–805
Richthammer C, Enseleit M, Sanchez-Leon E, Marz S, Heilig Y, Riquelme M, Seiler S (2012) RHO1 and RHO2 share partially overlapping functions in the regulation of cell wall integrity and hyphal polarity in *Neurospora crassa*. Mol Microbiol 85:716–733
Riedl J, Crevenna AH, Kessenbrock K, Yu JH, Neukirchen D, Bista M, Bradke F, Jenne D, Holak TA, Werb Z, Sixt M, Wedlich-Soldner R (2008) Lifeact: a versatile marker to visualize F-actin. Nat Methods 5:605–607
Rincon SA, Bhatia P, Bicho C, Guzman-Vendrell M, Fraisier V, Borek WE, Alves Fde L, Dingli F, Loew D, Rappsilber J, Sawin KE, Martin SG, Paoletti A (2014) Pom1 regulates the assembly of Cdr2-Mid1 cortical nodes for robust spatial control of cytokinesis. J Cell Biol 206:61–77
Riquelme M, Bartnicki-Garcia S (2004) Key differences between lateral and apical branching in hyphae of *Neurospora crassa*. Fungal Genet Biol 41:842–851
Riquelme M, Bartnicki-García S (2008) Advances in understanding hyphal morphogenesis: ontogeny, phylogeny and cellular localization of chitin synthases. Fungal Biol Rev 22:56–70
Riquelme M, Fischer R, Bartnicki-Garcia S (2003) Apical growth and mitosis are independent processes in *Aspergillus nidulans*. Protoplasma 222:211–215
Roberts-Galbraith RH, Gould KL (2008) Stepping into the ring: the SIN takes on contractile ring assembly. Genes Dev 22:3082–3088
Roberts-Galbraith RH, Ohi MD, Ballif BA, Chen JS, McLeod I, McDonald WH, Gygi SP, Yates JR III, Gould KL (2010) Dephosphorylation of F-BAR protein Cdc15 modulates its conformation and stimulates its scaffolding activity at the cell division site. Mol Cell 39:86–99
Rock JM, Lim D, Stach L, Ogrodowicz RW, Keck JM, Jones MH, Wong CC, Yates JR III, Winey M, Smerdon SJ, Yaffe MB, Amon A (2013) Activation of the yeast Hippo pathway by phosphorylation-dependent assembly of signaling complexes. Science 340:871–875
Rodrigo G, Elena SF (2011) Structural discrimination of robustness in transcriptional feedforward loops for pattern formation. PLoS One 6:e16904
Rosenberg JA, Tomlin GC, McDonald WH, Snydsman BE, Muller EG, Yates JR III, Gould KL (2006) Ppc89 links multiple proteins, including the septation initiation network, to the core of the fission yeast spindle-pole body. Mol Biol Cell 17:3793–3805
Saha S, Pollard TD (2012) Anillin-related protein Mid1p coordinates the assembly of the cytokinetic contractile ring in fission yeast. Mol Biol Cell 23:3982–3992
Samejima I, Miller VJ, Rincon SA, Sawin KE (2010) Fission yeast Mto1 regulates diversity of cytoplasmic microtubule organizing centers. Curr Biol 20:1959–1965
Sampson K, Lew RR, Heath IB (2003) Time series analysis demonstrates the absence of pulsatile hyphal growth. Microbiology 149:3111–3119
Sanchez-Diaz A, Marchesi V, Murray S, Jones R, Pereira G, Edmondson R, Allen T, Labib K (2008) Inn1

couples contraction of the actomyosin ring to membrane ingression during cytokinesis in budding yeast. Nat Cell Biol 10:395–406

Sanchez-Leon E, Riquelme M (2015) Live imaging of beta-1,3-glucan synthase FKS-1 in *Neurospora crassa* hyphae. Fungal Genet Biol 82:104–107

Saputo S, Kumar A, Krysan DJ (2014) Efg1 directly regulates ACE2 expression to mediate cross talk between the cAMP/PKA and RAM pathways during *Candida albicans* morphogenesis. Eukaryot Cell 13:1169–1180

Schmidt S, Hofmann K, Simanis V (1997) Sce3, a suppressor of the *Schizosaccharomyces pombe* septation mutant cdc11, encodes a putative RNA-binding protein. Nucleic Acids Res 25:3433–3439

Schmidt M, Bowers B, Varma A, Roh DH, Cabib E (2002) In budding yeast, contraction of the actomyosin ring and formation of the primary septum at cytokinesis depend on each other. J Cell Sci 115:293–302

Seiler S, Justa-Schuch D (2010) Conserved components, but distinct mechanisms for the placement and assembly of the cell division machinery in unicellular and filamentous ascomycetes. Mol Microbiol 78:1058–1076

Seiler S, Plamann M (2003) The genetic basis of cellular morphogenesis in the filamentous fungus *Neurospora crassa*. Mol Biol Cell 14:4352–4364

Seiler S, Vogt N, Ziv C, Gorovits R, Yarden O (2006) The STE20/germinal center kinase POD6 interacts with the NDR kinase COT1 and is involved in polar tip extension in *Neurospora crassa*. Mol Biol Cell 17:4080–4092

Shannon KB, Li R (2000) A myosin light chain mediates the localization of the budding yeast IQGAP-like protein during contractile ring formation. Curr Biol 10:727–730

Shen KF, Osmani AH, Govindaraghavan M, Osmani SA (2014) Mitotic regulation of fungal cell-to-cell connectivity through septal pores involves the NIMA kinase. Mol Biol Cell 25:763–775

Si H, Justa-Schuch D, Seiler S, Harris SD (2010) Regulation of septum formation by the Bud3-Rho4 GTPase module in *Aspergillus nidulans*. Genetics 185:165–176

Si H, Rittenour WR, Xu K, Nicksarlian M, Calvo AM, Harris SD (2012) Morphogenetic and developmental functions of the *Aspergillus nidulans* homologues of the yeast bud site selection proteins Bud4 and Axl2. Mol Microbiol 85:252–270

Singh NS, Shao N, McLean JR, Sevugan M, Ren L, Chew TG, Bimbo A, Sharma R, Tang X, Gould KL, Balasubramanian MK (2011) SIN-inhibitory phosphatase complex promotes Cdc11p dephosphorylation and propagates SIN asymmetry in fission yeast. Curr Biol 21:1968–1978

Snaith HA, Sawin KE (2003) Fission yeast mod5p regulates polarized growth through anchoring of tea1p at cell tips. Nature 423:647–651

Snaith HA, Samejima I, Sawin KE (2005) Multistep and multimode cortical anchoring of tea1p at cell tips in fission yeast. EMBO J 24:3690–3699

Song S, Grenfell TZ, Garfield S, Erikson RL, Lee KS (2000) Essential function of the polo box of Cdc5 in subcellular localization and induction of cytokinetic structures. Mol Cell Biol 20:286–298

Springer ML, Yanofsky C (1989) A morphological and genetic analysis of conidiophore development in *Neurospora crassa*. Genes Dev 3:559–571

Stajich JE, Berbee ML, Blackwell M, Hibbett DS, James TY, Spatafora JW, Taylor JW (2009) The fungi. Curr Biol 19:R840–R845

Straight AF, Cheung A, Limouze J, Chen I, Westwood NJ, Sellers JR, Mitchison TJ (2003) Dissecting temporal and spatial control of cytokinesis with a myosin II Inhibitor. Science 299:1743–1747

Sudbery PE (2001) The germ tubes of *Candida albicans* hyphae and pseudohyphae show different patterns of septin ring localization. Mol Microbiol 41:19–31

Sudbery PE (2011) Growth of *Candida albicans* hyphae. Nat Rev Microbiol 9:737–748

Sugawara T, Sato M, Takagi T, Kamasaki T, Ohno N, Osumi M (2003) In situ localization of cell wall alpha-1,3-glucan in the fission yeast *Schizosaccharomyces pombe*. J Electron Microsc (Tokyo) 52:237–242

Sun L, Guan R, Lee IJ, Liu Y, Chen M, Wang J, Wu JQ, Chen Z (2015) Mechanistic insights into the anchorage of the contractile ring by anillin and Mid1. Dev Cell 33:413–426

Takeshita N, Fischer R (2011) On the role of microtubules, cell end markers, and septal microtubule organizing centres on site selection for polar growth in *Aspergillus nidulans*. Fungal Biol 115:506–517

Takeshita N, Higashitsuji Y, Konzack S, Fischer R (2008) Apical sterol-rich membranes are essential for localizing cell end markers that determine growth directionality in the filamentous fungus *Aspergillus nidulans*. Mol Biol Cell 19:339–351

Takeshita N, Manck R, Grun N, de Vega SH, Fischer R (2014) Interdependence of the actin and the microtubule cytoskeleton during fungal growth. Curr Opin Microbiol 20:34–41

Tasto JJ, Morrell JL, Gould KL (2003) An anillin homologue, Mid2p, acts during fission yeast cytokinesis to organize the septin ring and promote cell separation. J Cell Biol 160:1093–1103

Tatebe H, Shimada K, Uzawa S, Morigasaki S, Shiozaki K (2005) Wsh3/Tea4 is a novel cell-end factor essential for bipolar distribution of Tea1 and protects cell polarity under environmental stress in *S. pombe*. Curr Biol 15:1006–1015

Teh EM, Chai CC, Yeong FM (2009) Retention of Chs2p in the ER requires N-terminal CDK1-phosphorylation sites. Cell Cycle 8:2964–2974

Teichert I, Steffens EK, Schnass N, Franzel B, Krisp C, Wolters DA, Kuck U (2014) PRO40 is a scaffold

protein of the cell wall integrity pathway, linking the MAP kinase module to the upstream activator protein kinase C. PLoS Genet 10:e1004582
Tran PT, Marsh L, Doye V, Inoue S, Chang F (2001) A mechanism for nuclear positioning in fission yeast based on microtubule pushing. J Cell Biol 153:397–411
Trinci AP, Collinge AJ (1973) Structure and plugging of septa of wild type and spreading colonial mutants of *Neurospora crassa*. Arch Mikrobiol 91:355–364
Ullal P, McDonald NA, Chen JS, Lo Presti L, Roberts-Galbraith RH, Gould KL, Martin SG (2015) The DYRK-family kinase Pom1 phosphorylates the F-BAR protein Cdc15 to prevent division at cell poles. J Cell Biol 211:653–668
Upadhyay S, Shaw BD (2008) The role of actin, fimbrin and endocytosis in growth of hyphae in *Aspergillus nidulans*. Mol Microbiol 68:690–705
Vargas-Muniz JM, Renshaw H, Richards AD, Lamoth F, Soderblom EJ, Moseley MA, Juvvadi PR, Steinbach WJ (2015) The *Aspergillus fumigatus* septins play pleiotropic roles in septation, conidiation, and cell wall stress, but are dispensable for virulence. Fungal Genet Biol 81:41–51
Vavylonis D, Wu JQ, Hao S, O'Shaughnessy B, Pollard TD (2008) Assembly mechanism of the contractile ring for cytokinesis by fission yeast. Science 319:97–100
Veith D, Scherr N, Efimov VP, Fischer R (2005) Role of the spindle-pole-body protein ApsB and the cortex protein ApsA in microtubule organization and nuclear migration in *Aspergillus nidulans*. J Cell Sci 118:3705–3716
Verdin J, Bartnicki-Garcia S, Riquelme M (2009) Functional stratification of the Spitzenkorper of *Neurospora crassa*. Mol Microbiol 74:1044–1053
Vinck A, de Bekker C, Ossin A, Ohm RA, de Vries RP, Wosten HA (2011) Heterogenic expression of genes encoding secreted proteins at the periphery of *Aspergillus niger* colonies. Environ Microbiol 13:216–225
Vogt N, Seiler S (2008) The RHO1-specific GTPase-activating protein LRG1 regulates polar tip growth in parallel to Ndr kinase signaling in Neurospora. Mol Biol Cell 19:4554–4569
Walker LA, Gow NA, Munro CA (2013) Elevated chitin content reduces the susceptibility of *Candida* species to caspofungin. Antimicrob Agents Chemother 57:146–154
Wang A, Raniga PP, Lane S, Lu Y, Liu H (2009) Hyphal chain formation in *Candida albicans*: Cdc28-Hgc1 phosphorylation of Efg1 represses cell separation genes. Mol Cell Biol 29:4406–4416
Wang N, Wang M, Zhu YH, Grosel TW, Sun D, Kudryashov DS, Wu JQ (2015) The Rho-GEF Gef3 interacts with the septin complex and activates the GTPase Rho4 during fission yeast cytokinesis. Mol Biol Cell 26:238–255
Watanabe S, Okawa K, Miki T, Sakamoto S, Morinaga T, Segawa K, Arakawa T, Kinoshita M, Ishizaki T, Narumiya S (2010) Rho and anillin-dependent control of mDia2 localization and function in cytokinesis. Mol Biol Cell 21:3193–3204
Weiss EL (2012) Mitotic exit and separation of mother and daughter cells. Genetics 192:1165–1202
Weiss EL, Kurischko C, Zhang C, Shokat K, Drubin DG, Luca FC (2002) The *Saccharomyces cerevisiae* Mob2p-Cbk1p kinase complex promotes polarized growth and acts with the mitotic exit network to facilitate daughter cell-specific localization of Ace2p transcription factor. J Cell Biol 158:885–900
Wendland J (2003) Analysis of the landmark protein Bud3 of *Ashbya gossypii* reveals a novel role in septum construction. EMBO Rep 4:200–204
Wendland J, Walther A (2005) *Ashbya gossypii*: a model for fungal developmental biology. Nat Rev Microbiol 3:421–429
Wightman R, Bates S, Amornrrattanapan P, Sudbery P (2004) In *Candida albicans*, the Nim1 kinases Gin4 and Hsl1 negatively regulate pseudohypha formation and Gin4 also controls septin organization. J Cell Biol 164:581–591
Wloka C, Bi E (2012) Mechanisms of cytokinesis in budding yeast. Cytoskeleton (Hoboken) 69:710–726
Wloka C, Nishihama R, Onishi M, Oh Y, Hanna J, Pringle JR, Krauss M, Bi E (2011) Evidence that a septin diffusion barrier is dispensable for cytokinesis in budding yeast. Biol Chem 392:813–829
Wolkow TD, Harris SD, Hamer JE (1996) Cytokinesis in *Aspergillus nidulans* is controlled by cell size, nuclear positioning and mitosis. J Cell Sci 109(Pt 8):2179–2188
Wosten HA, van Veluw GJ, de Bekker C, Krijgsheld P (2013) Heterogeneity in the mycelium: implications for the use of fungi as cell factories. Biotechnol Lett 35:1155–1164
Wu JQ, Sirotkin V, Kovar DR, Lord M, Beltzner CC, Kuhn JR, Pollard TD (2006) Assembly of the cytokinetic contractile ring from a broad band of nodes in fission yeast. J Cell Biol 174:391–402
Wu H, Guo J, Zhou YT, Gao XD (2015) The anillin-related region of Bud4 is the major functional determinant for Bud4's function in septin organization during bud growth and axial bud site selection in budding yeast. Eukaryot Cell 14:241–251
Xu Y, Moseley JB, Sagot I, Poy F, Pellman D, Goode BL, Eck MJ (2004) Crystal structures of a Formin Homology-2 domain reveal a tethered dimer architecture. Cell 116:711–723
Yarden O, Plamann M, Ebbole DJ, Yanofsky C (1992) cot-1, a gene required for hyphal elongation in *Neurospora crassa*, encodes a protein kinase. EMBO J 11:2159–2166
Yoshida S, Kono K, Lowery DM, Bartolini S, Yaffe MB, Ohya Y, Pellman D (2006) Polo-like kinase Cdc5

controls the local activation of Rho1 to promote cytokinesis. Science 313:108–111
Zekert N, Veith D, Fischer R (2010) Interaction of the *Aspergillus nidulans* microtubule-organizing center (MTOC) component ApsB with gamma-tubulin and evidence for a role of a subclass of peroxisomes in the formation of septal MTOCs. Eukaryot Cell 9:795–805
Zhang L, Maddox AS (2010) Anillin. Curr Biol 20:R135–R136
Zhang G, Kashimshetty R, Ng KE, Tan HB, Yeong FM (2006) Exit from mitosis triggers Chs2p transport from the endoplasmic reticulum to mother-daughter neck via the secretory pathway in budding yeast. J Cell Biol 174:207–220
Zhou M, Wang YL (2008) Distinct pathways for the early recruitment of myosin II and actin to the cytokinetic furrow. Mol Biol Cell 19:318–326
Ziv C, Kra-Oz G, Gorovits R, Marz S, Seiler S, Yarden O (2009) Cell elongation and branching are regulated by differential phosphorylation states of the nuclear Dbf2-related kinase COT1 in *Neurospora crassa*. Mol Microbiol 74:974–989

The Cytoskeleton and Polarity Markers During Polarized Growth of Filamentous Fungi

Norio Takeshita[1], Reinhard Fischer[2]

CONTENTS

I. Introduction

Maintaining cell polarity is essential for cells to ensure homeostasis and their proper functioning (Goehring and Grill 2013; Wu and Lew 2013). Symmetry breaking is often preceded by cytoskeleton-dependent polarization of certain key proteins as observed in epithelial cells with apical-basal polarity, neuronal differentiation from dendrites to axons, and migrating cells. Filamentous fungi are highly polarized eukaryotic cells, which continuously elongate their hyphae at the tips. Some distance back from the tip, hyphae can initiate new sites of polar growth in the process of branch formation. The establishment and maintenance of polar growth is one fascinating question in biology. Filamentous fungi are widely used as model systems for the analysis of the relationship between cell polarity and shape (Harris 2006; Fischer et al. 2008; Riquelme et al. 2011, 2018; Takeshita et al. 2014). Some filamentous fungi are pathogenic to animals and plants, and often growth in the host is accompanied by a change from hyphal growth to yeast-like growth or vice versa (dimorphism) (Garcia-Vidal et al. 2013). Other fungi are useful in biotechnology, such as enzyme production and fermentation in food industry due to their high ability of enzyme secretion (Punt et al. 2011). Thus, the analysis of polarized growth of filamentous fungi can contribute to the medical, agricultural, and biotechnological fields.

The filamentous ascomycete *Aspergillus nidulans* has been employed worldwide for more than 60 years as a model organism because it is closely related to clinically and economically important *Aspergilli* and it is easily manipulated in the laboratory. The most characteristic cell type of filamentous fungi is the vegetative hypha. This nonspecialized, syncytial (multinucleated) cell is characterized by a continuous polarized growth mode, mediated at the tip through the addition of new material that is transported from distal regions. In single-cell yeasts, such as in budding yeast *Saccharomyces cerevisiae* and in fission yeast *Schizosaccharomyces pombe*, polarized growth is restricted to certain times during the cell cycle, whereas in filamentous fungi cell extension is a continuous and indefinite process.

[1] Microbiology Research Center for Sustainability (MiCS), Faculty of Life and Environmental Sciences, University of Tsukuba, Tsukuba, Japan; e-mail: takeshita.norio.gf@u.tsukuba.ac.jp

[2] Department of Microbiology, Institute for Applied Biosciences, Karlsruhe Institute of Technology (KIT), Karlsruhe, Germany; e-mail: reinhard.fischer@kit.edu

Biology of the Fungal Cell, 3rd Edition
The Mycota VIII
D. Hoffmeister, M. Gressler (Eds.)

The extension of hyphal tips requires the continuous enlargement of the cell membrane and the extension of the cell wall. Both are achieved through massive vesicle fusion at the tip. The vesicles transport cell wall-synthesizing enzymes and provide new membrane. Vesicle transport as well as all other dynamic processes related to polar growth, such as organelle duplication and distribution, or the transport of RNA, proteins, or lipids, requires cytoskeletal elements. Many of those components were discovered early on in mutant screenings followed by genetic and later molecular biological analyses. For instance, tubulin was discovered through a combination of biochemical analyses with the analysis of mutants with altered sensitivity against the microtubule drug benomyl (*benA*) in *A. nidulans* (Oakley 2004). Mitotic elements were isolated as temperature-sensitive mutants with a *block-in-mitosis* (*bimA*, *bimC*, *bimE*) or mutants, which never entered mitosis (*never-in-mitosis*, *nimA*, *nimX*, *nimT*) (Orr and Rosenberger 1976; Morris and Enos 1992; Osmani and Mirabito 2004). In other mutants of this screening, *nuclear distribution* (*nudA*, *nudE*) was affected (Meyer et al. 1988). In subsequent suppressor screenings using the *benA33* mutant, *tubA* and *mipA* were discovered (Morris et al. 1979; Oakley and Oakley 1989). The genes encoded beta tubulin (*benA*), alpha tubulin (*tubA*), or gamma tubulin (*mipA*), dynein (*nudA*), or kinesin (*bimC*). The mutagenesis approaches performed in *S. cerevisiae* and *S. pombe* but also filamentous fungi such as *A. nidulans* revealed a wealth of information, which could probably not have generated by other means. The improved molecular biological methods and the genome information opened the possibility of reverse-genetic approaches. With these approaches the role of proteins of conserved pathways was studied in other organisms and organism-specific functions were discovered. However, recent major advances in cost-effective sequencing of entire genomes have the great potential to revolutionize our approaches again and allow intelligent mutant screening followed by bulk sequencing of mutant genomes. This strategy reduced mutant analysis from months or years to weeks or months (Tan et al. 2014).

Polarized growth is thus studied by genetic, molecular biological, biochemical, and cell biological methods. However, this research field has benefited more than others from the combination of the still ongoing improvement of the microscopic techniques and the development of fluorescent reporter proteins in recent years. Several overviews have summarized different aspects of polarized growth in fungi (Chang and Peter 2003; Harris and Momany 2004; Penalva 2010; Berepiki et al. 2011; Steinberg 2011, 2014; Sudbery 2011; Egan et al. 2012a).

Here we review recent findings unraveling the mechanism of polarized growth with special emphasis on the roles of the actin and microtubule (MT) cytoskeletons, polarity markers linking the two cytoskeletons.

II. The Actin Cytoskeleton

The actin cytoskeleton plays a central role in cell morphology of eukaryotic cells (Dominguez and Holmes 2011). **Actin filaments** (**F-actin**), which are composed of linear polymers of **G-actin** subunits, generate force against the plasma membrane and also act as tracks for myosin motors. The dynamic cycles of polymerization and depolymerization of G-actin and F-actin are involved in many different key cellular processes, such as cell motility, cytokinesis, secretion, and the control of cell morphology (Michelot and Drubin 2011).

There are three high-order **F-actin** structures with distinct functions in filamentous fungi: actin rings, patches, and cables (Berepiki et al. 2011). Studies using anti-actin chemical agents confirmed that a polymerized actin cytoskeleton is required for normal apical growth and hyphal tip shape in different fungal organisms (Torralba et al. 1998). Phalloidin conjugated to fluorescent dyes has been widely used for imaging F-actin in eukaryotic cells including fungi such as budding yeast (Amberg 1998), fission yeast (Pelham and Chang 2001), and *Ashbya gossypii* (Walther and Wendland 2004) but does not work in most filamentous fungi (Brent Heath et al. 2003). The immunostaining of actin using an anti-actin antibody and GFP-labeled actin or an actin-binding protein (AbpA in *A. nidulans*) revealed the localization of the F-actin structures, mainly actin rings and actin patches, in filamentous fungi (Araujo-Bazan et al. 2008). However, visualization of actin cables was difficult by these methods. Recently, specific markers for actin cables, such as Lifeact and tropomyosin, were developed to visualize them (Taheri-Talesh et al. 2008; Berepiki et al. 2010; Delgado-Alvarez et al. 2010). Lifeact, which consists of 17 amino acids from the N-terminus of Abp140p of *S. cerevisiae*, has been shown to be a marker for F-actin binding and labeling in vitro and in yeast cells (Asakura et al. 1998; Riedl et al. 2008).

The **actin rings** in cooperation with class II myosin function in septum formation (Taheri-Talesh et al. 2012; Delgado-Alvarez et al. 2014) (Fig. 1a). Septum formation proceeds according to the following series of steps: actin and myosin tangle assembly at a septation site, contraction of the actin ring (or called actomyosin ring), actin-mediated invagination of the plasma membrane, and deposition of the chitinous primary septum (Momany and Hamer 1997; Delgado-Alvarez et al. 2014). Mutant analysis of class II myosin (*myoB* mutation in the converter subdomain in *A. nidulans*) suggests that the motor activity is necessary for the contraction of the actin ring (Hill et al. 2015). The actin ring assembly and septum formation are controlled through the nuclear position and cell cycle progression in *A. nidulans* (Harris 2001).

Actin patches are peripheral punctate structures where probably the endocytic machinery localizes (Araujo-Bazan et al. 2008) (Fig. 1b). The predominant localization of these patches at subapical regions suggests spatial coupling of apical exocytosis and subapical compensatory endocytosis (Penalva 2010). Class I myosins function in

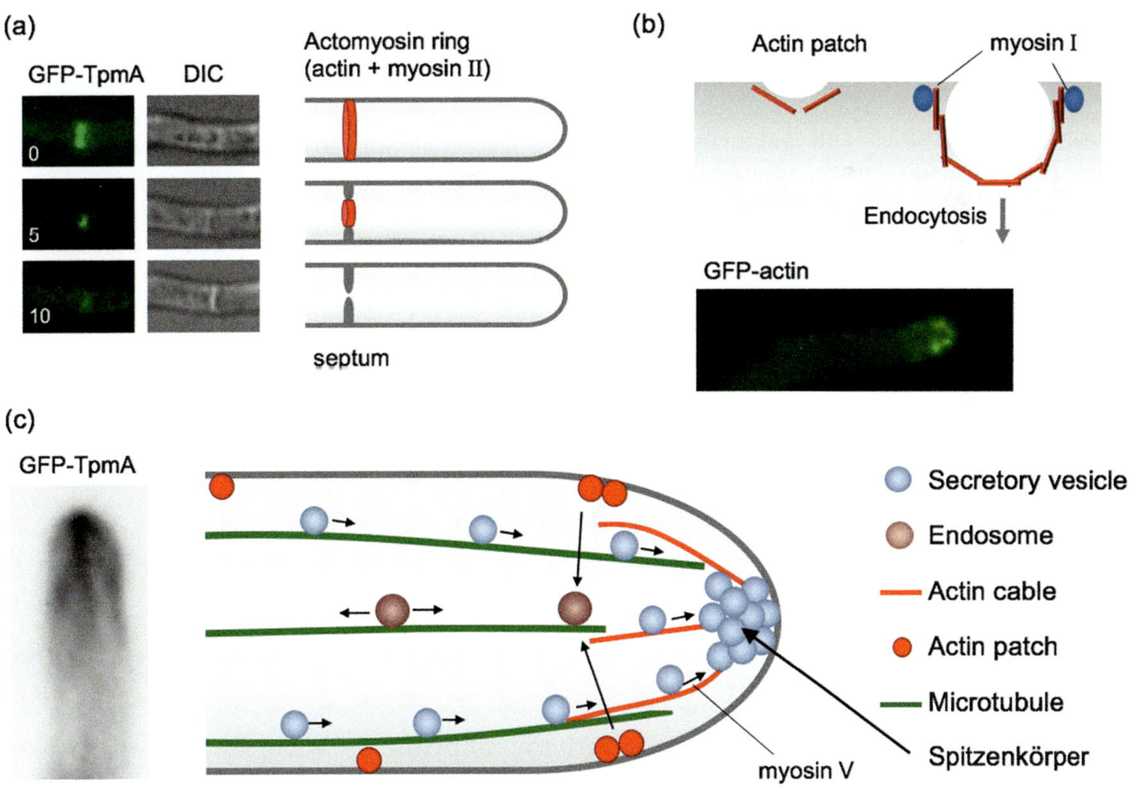

Fig. 1 Distinct roles of actin cytoskeletons and related myosin in *A. nidulans*. (**a**) Actin ring and myosin II (actomyosin ring) for septation. (**b**) Actin patch and myosin I for endocytosis. (**c**) Actin cables and myosin V for exocytosis. Scheme of tip growth in *A. nidulans* hyphae. Vesicle trafficking via the actin and MT cytoskeleton. Before fusion with the plasma membrane, secretion vesicles accumulate at Spitzenkörper

endocytosis to support invagination of endocytic vesicles (Kaksonen et al. 2006). Mutant analysis of *myoA*, the class I myosin in *A. nidulans*, revealed that the function of MyoA in endocytosis is regulated through phosphorylation by a member of the p21-activated kinase (PAK) family (Yamashita and May 1998). The mutant phenotype of genes involved in endocytosis indicates that endocytosis is essential (Araujo-Bazan et al. 2008). Besides the internalization of extracellular molecules, plasma membrane proteins, and lipids by endocytosis, endocytic recycling of polarized material at the hyphal tip and a balance between endocytosis and exocytosis at the hyphal tip are assumed to control polarized growth and cell shape (Shaw et al. 2011).

Actin cables are linear bundles of short actin filaments nucleated by formins that are present at the apex of the hyphae. As mentioned before, dynamic actin cables are generally very difficult to visualize; however, recently specific markers, such as Lifeact and tropomyosin, were developed (Taheri-Talesh et al. 2008; Berepiki et al. 2010; Delgado-Alvarez et al. 2010). In *N. crassa*, Lifeact has been used to visualize dynamic actin cables and patches (Berepiki et al. 2010; Delgado-Alvarez et al. 2010). However, it has to be considered that overexpression of the construct may cause some artifacts (Bergs et al. 2016). Tropomyosin is a conserved actin filament-binding protein and regulates the interaction between actin and myosin in response to Ca^{2+} (Gunning et al. 2005). Tropomyosin has been used as a marker for actin cables in *A. nidulans* and *N. crassa* (Evangelista et al. 2002; Pearson et al. 2004; Taheri-Talesh et al. 2008). GFP-labeled tropomyosin, TpmA in *A. nidulans*, revealed the dynamic behavior with cycles of elongation and shrinkage (Fig. 1c) (Bergs et al. 2016). Multiple actin cables were formed from the hyphal tip with each actin cable showing elongation and shrinkage in an independent manner.

Actin cables are present at the apex of hyphae and are thought to serve as tracks for class V myosin-dependent secretory vesicle transport to the tip (Fig. 1c) (Taheri-Talesh et al. 2008, 2012; Berepiki et al. 2011; Pantazopoulou et al. 2014). The "basic" growth machinery involved in the formation of actin cables, vesicle transport, and exocytosis, such as formin, the polarisome (protein complex Spa2, Pea2, Aip3/Bud6, and formin Bni1 in *S. cerevisiae*), class V myosin, and the exocyst complex (octameric protein complex involved in the tethering of post-Golgi vesicles to the plasma membrane prior to vesicle fusion), is relatively conserved among eukaryotic cells and localized to the apex of hyphae (Harris et al. 2005; Sudbery 2011). Maturation of late Golgi cisternae into exocytic post-Golgi carriers was visualized in *A. nidulans* (Pantazopoulou et al. 2014). These carriers move on a MT-based bidirectional conveyor belt relaying them to actin, which ultimately focuses exocytosis at the apex.

III. Spitzenkörper

Before fusion, the secretion vesicles accumulate at the hyphal tip in a structure called **Spitzenkörper** or vesicle supply center (VSC) (Grove and Bracker 1970; Harris et al. 2005), a special structure in filamentous fungi, which determines growth direction of the hyphae (Bartnicki-Garcia et al. 1995; Riquelme and Sanchez-Leon 2014; Riquelme et al. 1998, 2014) (Fig. 1c). A VSC in motion provides a rational basis to predict how the secretory apparatus generates morphogenesis. The Spitzenkörper is believed to function as a VSC that regulates the delivery of cell wall-building vesicles to the apical cell surface, since the simulation analysis of VSC advance as a Spitzenkörper enables to mimic the hyphal growth of *N. crassa* (Bartnicki-García et al. 1989; Riquelme et al. 2000). The exact composition and organization of the Spitzenkörper has been vigorously elucidated (Riquelme et al. 2007, 2014; Sanchez-Leon et al. 2011; Fajardo-Somera et al. 2015). In *N. crassa*, all chitin synthases localize at the Spitzenkörper core, whereas macrovesicles carrying a β-1,3-glucan synthase complex occupy the Spitzenkörper outer layer. Similar spatial distribution of DnfA and DnfB, two P4 ATPases, in the Spitzenkörper was observed in *A. nidulans* (Schultzhaus et al. 2015).

IV. Septins

As filament-forming proteins, **septins** can be considered as a part of the cytoskeleton. Septins are a conserved family of GTP-binding proteins that form filaments in fungi and animals. Different septins form protein complexes with each other and form heteropolymers showing a variety of higher-order structures (Mostowy and Cossart 2012). In *S. cerevisiae*, five septin proteins (Cdc3, Cdc10, Cdc11, Cdc12, Shs1) form an hourglass structure associated with the plasma membrane at the mother-bud neck (Bertin et al. 2012). They act as a scaffold for proteins involved in cell division (Oh and Bi 2011). In filamentous fungi, in addition to septum formation, septins have been shown to have roles in morphogenesis, coordinating nuclear division, and organizing the cytoskeleton (Lindsey and Momany 2006; Gladfelter 2010; Bridges and Gladfelter 2014).

In the hyphal form of the human pathogen *Candida albicans*, septins form three distinct structures, a diffuse band at the base of hyphae, a bright double ring at septation sites, and a diffuse cap at hyphal tips (Sudbery 2001; Warenda and Konopka 2002). In the filamentous phytopathogen *A. gossypii*, septins form a diffuse cap at hyphal tips and rings composed of discrete bars at septation sites and at newly emerging branches (Helfer and Gladfelter 2006; DeMay et al. 2009). In *C. albicans* and *A. gossypii*, the diffuse cap initially appears at the hyphal tip and travels with the hyphal tip until an unknown signal triggers detachment from the tip and the formation of an anchored, higher-order ring. The higher-order ring encircles the hyphal cell cortex and persists at that location, while the tip then continues to grow. Septation sites are determined by positions of the high-order ring that detaches from the tip and anchors at the hyphal cell cortex in these fungi. The structural change of septins is regulated through their phosphorylation by cyclin-dependent kinases in *C. albicans* (Sinha et al. 2007) or septin-associated kinases in *A. gossypii* (DeMay et al. 2009).

In *A. nidulans* and *N. crassa*, septins form different higher-order structures, rings, filaments, bars, bands, and caps (Westfall and Momany 2002; Lindsey et al. 2010; Berepiki and Read 2013; Hernandez-Rodriguez et al. 2014). They are important for septation, germination, branch emergence, and asexual spore formation. In the basidiomycete plant pathogen *Ustilago maydis*, septins localize to a variety of structures, collars at the bud neck and filaments at growing cell tips that run along the length of the cell and partially colocalize with MTs (Boyce et al. 2005; Alvarez-Tabares and Perez-Martin 2010). Septin filaments in the basidiomycete dikaryotic hyphae of *Cryptococcus neoformans* have also been shown to occasionally colocalize with MTs (Kozubowski and Heitman 2010). These septins have roles in morphogenesis and host infection. In the plant pathogen *Magnaporthe oryzae*, the location of the appressorium septum is determined by the site of septin ring assembly (Saunders et al. 2010). The septin ring functions as a scaffold for proteins required for appressorium formation (Dagdas et al. 2012). The recent reports on septins in filamentous fungi have revealed new roles for these cytoskeletal polymers.

V. The Microtubule Cytoskeleton

MTs play a crucial role during mitosis but serve also additional functions in interphase in filamentous fungi. They are important for the distribution of nuclei and other organelles and serve as tracks for endosomes and other vesicles; thus they are important for rapid hyphal growth (Xiang and Fischer 2004; Horio and Oakley 2005; Egan et al. 2012a; Steinberg 2014).

The rather stable minus end of MTs is located at the **MT-organizing center (MTOC)**, whereas the plus end is facing to the cell periphery with alternate growing and shrinking phases. In filamentous fungi, spindle pole bodies (SPBs) serve as MTOCs (Fig. 2a) (Oakley et al. 1990). They contain γ-tubulin, first discovered in *A. nidulans*, which is required for nucleation of MTs (Oakley and Oakley 1989; Oakley et al. 1990). Furthermore, there is good evidence that areas close to the septa act as MTOCs in *A. nidulans* (**sMTOCs**) (Fig. 2a) (Veith et al. 2005; Xiong and Oakley 2009; Zekert and Fischer 2009; Zhang et al. 2017). The composition of those MTOCs, their role, and their tethering to the septal membrane remain to be elusive. However, there is also good evidence for several MTOCs in *S. pombe*, in *U. maydis*, and in higher eukaryotes. In *S. pombe* equatorial (**eMTOC**) and interphase MTOCs (**iMTOC**) have been described (Sawin and Tran 2006) but appear to be only temporally active during certain stages of the cell cycle, and their exact nature also is enigmatic. In higher eukaryotes the Golgi apparatus had some MT-forming activity (Sutterlin and Colanzi 2010).

In the tip compartment of *A. nidulans*, most MTs are oriented with their dynamic plus ends toward the hyphal tip (Fig. 2a, b) (Konzack et al. 2005). There are only a few MTs found in interphase compartments, and nuclei migrate probably along MTs until they reach a certain

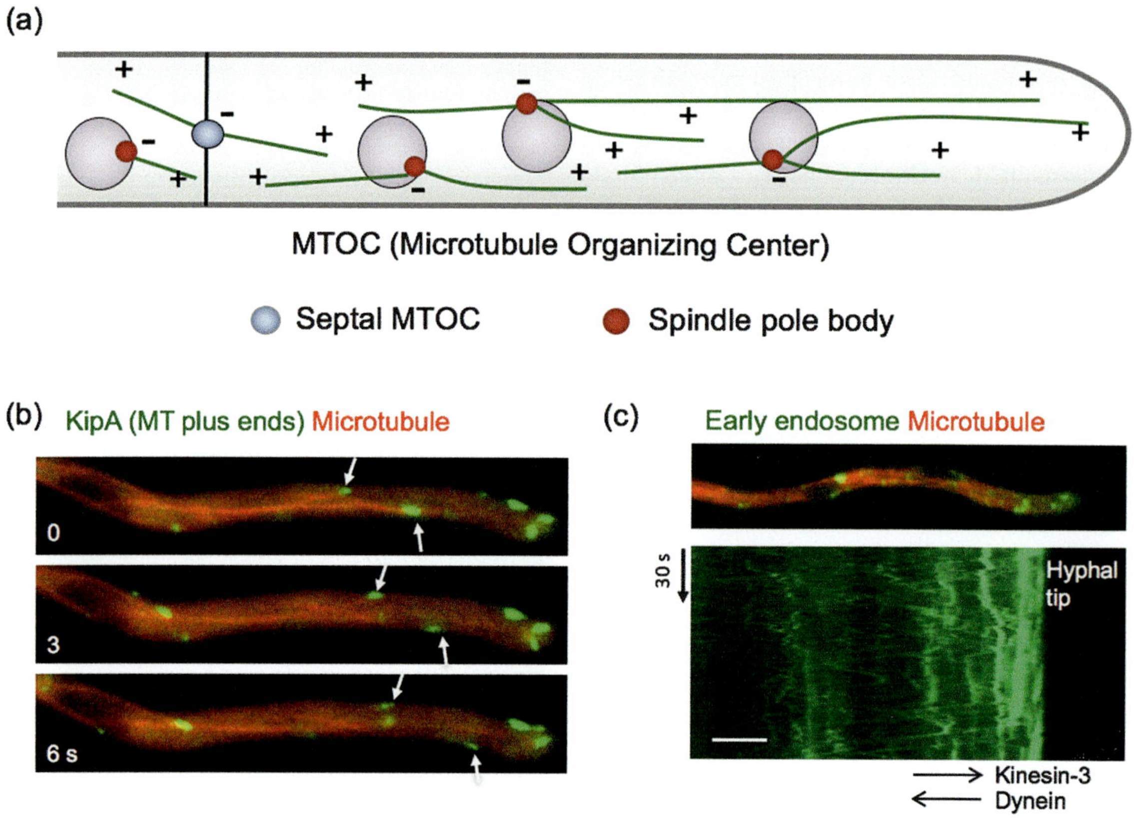

Fig. 2 Microtubule cytoskeleton in *A. nidulans*. (**a**) MT-organizing center (MTOC) at spindle pole bodies and septa. Most MTs are oriented with their dynamic plus ends toward the hyphal tip in the tip compartment. (**b**) Time lapse of GFP-KipA (MT plus end marker) and mCherry-TubA (MT) in *A. nidulans* hyphae. (**c**) Image of GFP-RabA (early endosome marker) and mCherry-TubA (MT) and kymograph of GFP-RabA

position. The entire hypha looks therefore very organized with evenly spaced nuclei.

Two classes of MT-dependent motors, the minus end-directed **dynein** and the plus end-directed **kinesins**, are involved in positioning of organelles and transport of membranes. Whereas genomes of filamentous fungi contain a single dynein motor, they usually encode 10–12 kinesins (Schoch et al. 2003). The function of kinesin-3 and the dynein motor in the transport of early endosomes have been analyzed deeply (Fig. 2c) (Steinberg 2011, 2014; Egan et al. 2012a). In *A. nidulans* hyphae, most MTs between the hyphal tip and the most proximal nucleus were polarized with their plus ends oriented toward the growing hyphal tip. Dynein and kinesin-3/UncA are the opposite motors responsible for bidirectional transport of endosomes and peroxisomes (Abenza et al. 2009; Egan et al. 2012b). Studies of *A. nidulans* kinesin-3 implicated indirect evidence for the existence of a subpopulation of detyrosinated MTs (Zekert and Fischer 2009; Seidel et al. 2012). However, a final proof for the existence of posttranslationally modified tubulin in fungi is yet still missing.

Endocytic recycling at subapical regions supports fungal tip growth. The bidirectional motility of early endosomes is thought to be involved in sorting the endocytic cargo to the subapical vacuole for degradation (Wedlich-Soldner et al. 2000). An unexpected new role for early endosome motility was revealed in *U. maydis*. The RNA-binding protein Rrm4 transports various mRNAs on moving early endosomes, suggesting that early endosome motility toward the cell ends supports polar delivery of Rrm4-bound RNAs (Konig et al. 2009; Baumann et al. 2012). In addition, it was shown that the mRNA of the septins *cdc3* and *cdc12* is actively translated on moving early endosomes, and both proteins bind to early endosomes (Baumann et al. 2014). In fact, it was confirmed that ribosomes attach to early endosomes and are translationally active during the transport (Higuchi et al. 2014). It is suggested that bidirectional early endosome motility constantly stirs the translation machinery, which diffuses passively in the cytoplasm and thereby contributes to distribute the ribosomes throughout the cell (Steinberg 2014; Haag et al. 2015). Indeed, endosomal transport of heteromeric septin complexes along microtubules is crucial for formation of higher-order structures in *U. maydis* (Zander et al. 2016).

The deletion of **conventional kinesin (kinesin-1)** decreased the growth rate and caused defects in Spitzenkörper stability, protein secretion, and pathogenicity (Lehmler et al. 1997; Seiler et al. 1997, 1999; Requena et al. 2001; Schuster et al. 2012). These results suggest a possible conserved role in vesicle transportation similar to higher eukaryotic cells. High-speed imaging revealed vesicle transport of chitin synthases (Takeshita et al. 2015). The frequency of the transport was clearly decreased in the absence of *kinA*, kinesin-1 in *A. nidulans*. Secretory vesicles are thought to be transported by kinesin-1 along MTs for long distances toward hyphal tips in filamentous fungi, although the localization of the ER and the Golgi close to hyphal tips raises questions about the function and cargo of kinesin-1 (Markina-Inarrairaegui et al. 2013; Pinar et al. 2013). Possibly long-distance transport of secretion vesicles is less important and that actin-dependent movement is rather sufficient. Indeed, hyphal extension can occur quite long without functional MTs but is immediately stopped as soon as the integrity of the actin cytoskeleton is disturbed (Torralba et al. 1998; Horio and Oakley 2005). Although the role of MTs and the different cytoskeletons could be diverse in different fungi, vesicle movement and delivery to the tip plasma membrane likely depend on the cooperation of actin- and MT-dependent motors (Zhang et al. 2011; Schuster et al. 2012; Taheri-Talesh et al. 2012).

VI. Cell-End Markers for Polarity Maintenance

Cell polarity is essential for the proper functioning of many cell types. During cellular morphogenesis—from fission yeast to human cells—MTs deliver positional information to the proper site of cortical polarity (Siegrist and Doe 2007; Li and Gundersen 2008). The polarization of the actin cytoskeleton and signal transduction cascades and continuous membrane transport toward the growth site depend on MTs.

Because MT dynamics and many MT functions are conserved among eukaryotes, lower eukaryotes can serve as excellent models. In ***S. pombe***, the **kelch-repeat protein Tea1** is delivered by growing MTs to the cell ends (Fig. 3a) (Mata and Nurse 1997). Tea1 reaches the MT plus end with the kinesin-7, **Tea2** (Browning et al. 2000, 2003), and is anchored at the cell end membrane through the interaction with the prenylated protein, **Mod5** (Snaith and Sawin 2003). At the cell end, Tea1 interacts with additional components, which ultimately recruit the **formin For3** (Martin and Chang 2003; Feierbach et al. 2004; Tatebe et al. 2008). For3 forms actin cables required for exocytosis and polarized growth. **Cell-end markers**, Tea1 and Mod5, thus transmit positional information regulated by MTs to the actin cytoskeleton and thereby contribute to polarized growth. The cell-end marker genes were identified after analysis of polarity mutants (T-shaped or bent cells). The mutants of *tea1* and *tea2* (tip elongation aberrant) and *mod5* (morphology defective) exhibit T-shaped or bent cells as a result of the mislocalization of the polarity site away from the center of the cell end.

The molecular mechanism of MTs to regulate polarity maintenance is principally conserved from *S. pombe* to filamentous fungi (Fig. 3a) (Riquelme et al. 1998; Fischer et al. 2008; Takeshita et al. 2008; Higashitsuji et al. 2009; Takeshita and Fischer 2011). Tea1 and Tea2 are conserved in ***A. nidulans*** as **TeaA** (Tea1) and **KipA** (Tea2), respectively (Konzack et al. 2005; Takeshita et al. 2008). Although the Mod5 sequence is not conserved in filamentous fungi, a functional counterpart, **TeaR**, was discovered by screening for proteins that harbor a C-terminal prenylation motif in *A. nidulans* (Takeshita et al. 2008). TeaA is conserved in *Ascomycetes* and some *Basidiomycetes*, such as *Cryptococcus neoformans* and *Puccinia graminis*,

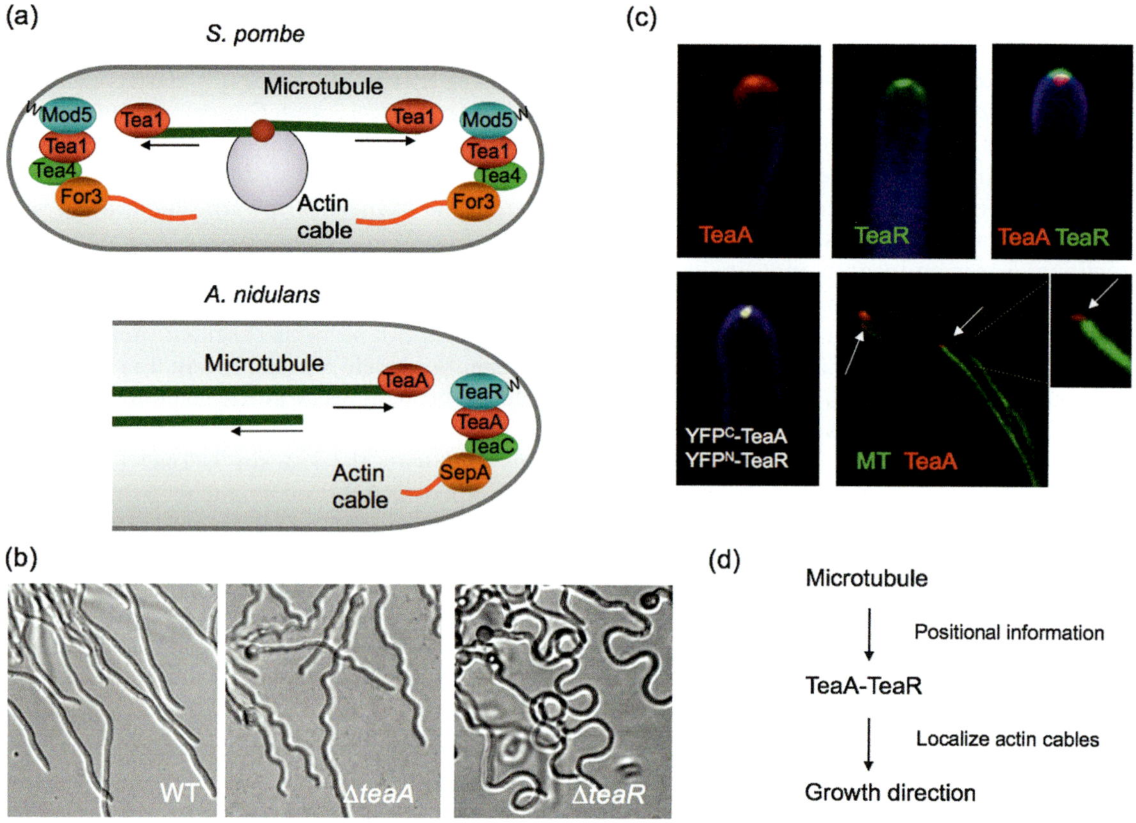

Fig. 3 Role of cell-end markers in *A. nidulans*. (**a**) Scheme of the function of cell-end markers in *S. pombe* and *A. nidulans*. (**b**) Differential interference contrast images of wild type, Δ*teaA*, Δ*teaR* strains. Δ*teaR* strains exhibited curved hyphae and Δ*teaA* strains exhibited zigzag hyphae. (**c**) Localization of mRFP1-TeaA and GFP-TeaR (upper) at the hyphal tips. Bimolecular fluorescence complementation (BiFC) assay of TeaA and TeaR at the hyphal tip (lower left). Localization of mRFP1-TeaA at the hyphal tip and MT plus end (lower right). (**d**) Scheme of growth direction regulated by cell-end markers and cell cytoskeletons

but not in *Zygomycetes*. TeaR is generally conserved in *Ascomycetes* except in *Hemiascomycetes*. The *A. nidulans kipA*, *teaA*, and *teaR* mutants show defects in polarity maintenance, which leads to curved (Δ*kipA* or Δ*teaR*) or zigzag growing hyphae (Δ*teaA*) (Fig. 3b). The two cell-end markers, TeaA and TeaR, localize at hyphal tips interdependently (Fig. 3c). TeaA is delivered to hyphal tips by growing MTs (Takeshita and Fischer 2011) and anchored to the hyphal tip cortex though the interaction with TeaR (Takeshita et al. 2008). TeaA, along with TeaC, recruits the formin, SepA, to the growth zone (Higashitsuji et al. 2009). The conserved mechanism of "cell-end markers," TeaA and TeaR, to transmit positional information from MTs to the actin cytoskeleton is required for maintenance of polarity and the growth direction of hyphae (Fig. 3d). Although the cell-end marker proteins appear to be conserved in other filamentous fungi, it remains to be studied if the roles assigned to them in *A. nidulans* are also conserved. For instance, the MT cytoskeleton is much more complex in *N. crassa*.

MTs in *A. nidulans* are necessary to maintain the polarity at the tip of hyphae through cell-end markers. Besides this function, MTs have additional functions in filamentous fungi, such as nuclear distribution and the movement of vesicles and other organelles; thus they are important for rapid hyphal growth. In contrast, MTs in *S. pombe* are required for polarity maintenance through cell-end markers but are not necessary for vesicle trafficking. Actin cables grow toward the growing cell ends, and Myo52, a myosin V, is responsible for polarized secretion of vesicles along actin cables and hence membrane enlargement and secretion of cell wall-synthesizing enzymes (Montegi et al. 2001; Win et al. 2001; Mulvihill et al. 2006).

VII. Rho GTPase

The small Rho-type GTPases Cdc42 and Rac1 are key regulators of eukaryotic cell polarity (Jaffe and Hall 2005; Virag et al. 2007). The switching between inactive GDP-bound and active GTP-bound states is controlled by guanine nucleotide exchange factors (GEFs) (Schmidt and Hall 2002) and GTPase-activating proteins (GAPs) (Bernards and Settleman 2004). Active GTPases stimulate multiple effector molecules, such as p21-activated kinases (PAKs), mitogen-activated protein kinases (MAPKs), formin, and subunits of the exocyst complex, which regulate numerous cellular processes including the rearrangement of the actin cytoskeleton, targeted vesicle transport, and exocytosis (Bishop and Hall 2000). In *S. cerevisiae*, two positive feedback loops are thought to contribute to Cdc42 polarization. One pathway involves recruitment of GEFs and effector complexes from the cytoplasm in a cytoskeleton-independent manner (Johnson et al. 2011). The other one is a vesicle-recycling feedback loop, where Cdc42 orients actin cables, which in turn deliver Cdc42 as cargo on secretory vesicles (Wedlich-Soldner et al. 2003).

In *S. cerevisiae* and *S. pombe*, Rac1 orthologs are not conserved, and Cdc42 alone is necessary and sufficient to control polarized growth (Adams et al. 1990; Miller and Johnson 1994). In contrast, filamentous fungi require both Rho GTPases to regulate their hyphal growth (Mahlert et al. 2006; Virag et al. 2007; Lichius et al. 2014). Both Cdc42 and Rac1 share at least one overlapping function that is required for polarity establishment. The combination of Δ*cdc42* with Δ*rac1* appeared synthetically lethal in *A. nidulans* (Virag et al. 2007). The cell-end marker deletion strains in *A. nidulans* lose the axis of polarity, although the hyphae still continue polarized tip growth. Although the polarization of Cdc42 and Rac1 by positive feedback loops as known in *S. cerevisiae* remains elusive in filamentous fungi, the Rho GTPases are possibly involved in polarity establishment independently of the cell-end markers in *A. nidulans*.

VIII. Cell-End Markers for Polarity Establishment

MTs and cell-end markers are not essential for spore germination itself but only for **site selection of germination** (Takeshita and Fischer 2011). During conidia germination, TeaA appeared at growth sites prior to the appearance of a small germination bud, and then MTs contact the cortex of the tip, while the polarized MTs likely deliver more TeaA and other proteins to the bud site, enforcing polar growth. Wild-type spores always germinate at only one site to form one hypha; in contrast, spores of Δ*teaA* often germinated at two (40%) or three sites (3%) simultaneously (Fig. 4a). A similar phenotype of multi-germtube formation was observed using the MT-destabilizing drug benomyl. It had been shown already that MTs are not essential for the germination process itself (Oakley and Morris 1980). TeaA and MTs are not necessary for the emergence of the germtube but probably rather for restricting germination to a specific place.

When the spores start germination, they grow isotropically at first, and then they switch to polarized growth with new material added to the tip of an emerging germtube. In *A. nidulans*, temperature-sensitive ***swo* (swollen cell) mutants** representing genes involved in polarity establishment and polarity maintenance were isolated (Momany et al. 1999). *swoC* (putative pseudouridylate synthase) and *swoF* (N-myristoyl transferase) (Shaw et al. 2002) are required to establish polarity, while *swoA* (O-mannosyltransferase) is required to maintain polarity. Besides that, the terminal phenotype of Δ*cdc42*Δ*rac1* indicates their functions are required to establish polarity (Virag et al. 2007). In addition, the morphological abnormalities of mutants involved in endocytosis indicate the importance of endocytosis in polarity establishment and polarity maintenance (Lee et al. 2008; Hervas-Aguilar and Penalva 2010; Shaw et al. 2011).

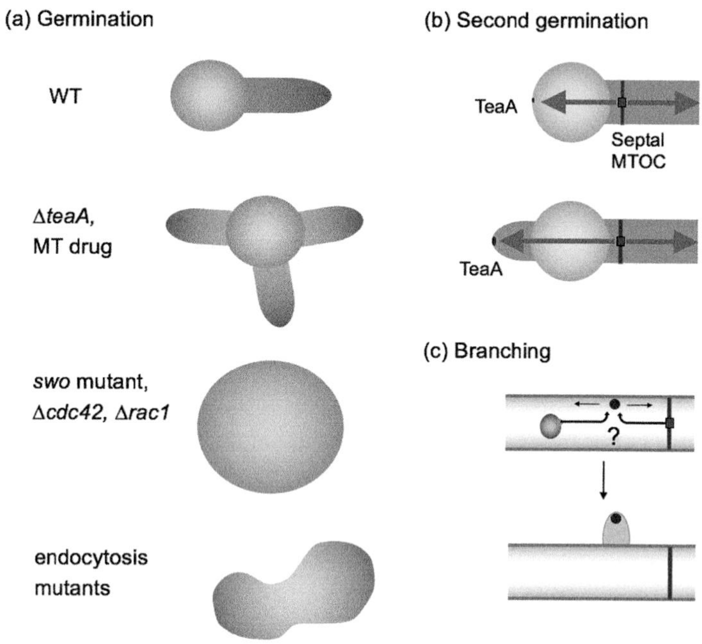

Fig. 4 Cell-end markers for polarity establishment. (**a**) Germination in wild type, and several mutants. (**b**) Scheme of second germination site selection by MTs from septal MTOC and TeaA. (**c**) Branching site selection

Once the first hypha reaches a determinate length, a second germtube appears on the spore. This **second germination site** normally lies opposite of the first hypha (Harris et al. 1999). The second germtube appeared after the first septum at the base of the first hypha was formed. TeaA appeared at the second germination site, opposite the first hypha, after septation in the first hypha. In *A. nidulans*, MTs are formed from SPBs and from septal MTOCs (Konzack et al. 2005; Veith et al. 2005; Xiong and Oakley 2009). MTs emanating from the septum of the first hypha grew toward the first germtube as well as in the direction of the spore and reached the second germination tip where TeaA localized (Fig. 4b). MTs originating from the septal MTOCs are thus important for TeaA delivery and may explain the bipolar germination pattern (Takeshita and Fischer 2011).

There appears to be two patterns of **lateral branching:** branches associated with septa and random branching. In several fungi including members of the *Saccharomycotina* (*A. gossypii*), zygomycetes (*Basidiobolus ranarum*), and basidiomycetes (*Coprinus* species), new branches emerge adjacent to septa (Trinci 1974). A random pattern of branching is observed in *A. nidulans*. Although the cell-end markers localize to branching sites prior to branching emerges, the mechanism of branch site selection remains largely unknown (Fig. 4c) (Harris 2008).

IX. Cell-End Markers for Polarity Focusing

The role of MTs in transmitting **positional information** through delivery of cell-end markers to the growth machinery is conserved in both *S. pombe* and *A. nidulans*. A Spitzenkörper, however, can only be observed in filamentous fungi but not at cell ends of fission yeast (Fig. 5a). This difference could be due to different growth speeds (Kohli et al. 2008). Another possible reason is that the cell-end markers concentrate at the apex of hyphae in *A. nidulans*, whereas the cell-end markers localize at multiple sites along cell ends in fission yeast (Dodgson et al. 2013). The positive feedback loop defined through the interdependence of TeaA and TeaR could be important for their concentration, but not sufficient because this

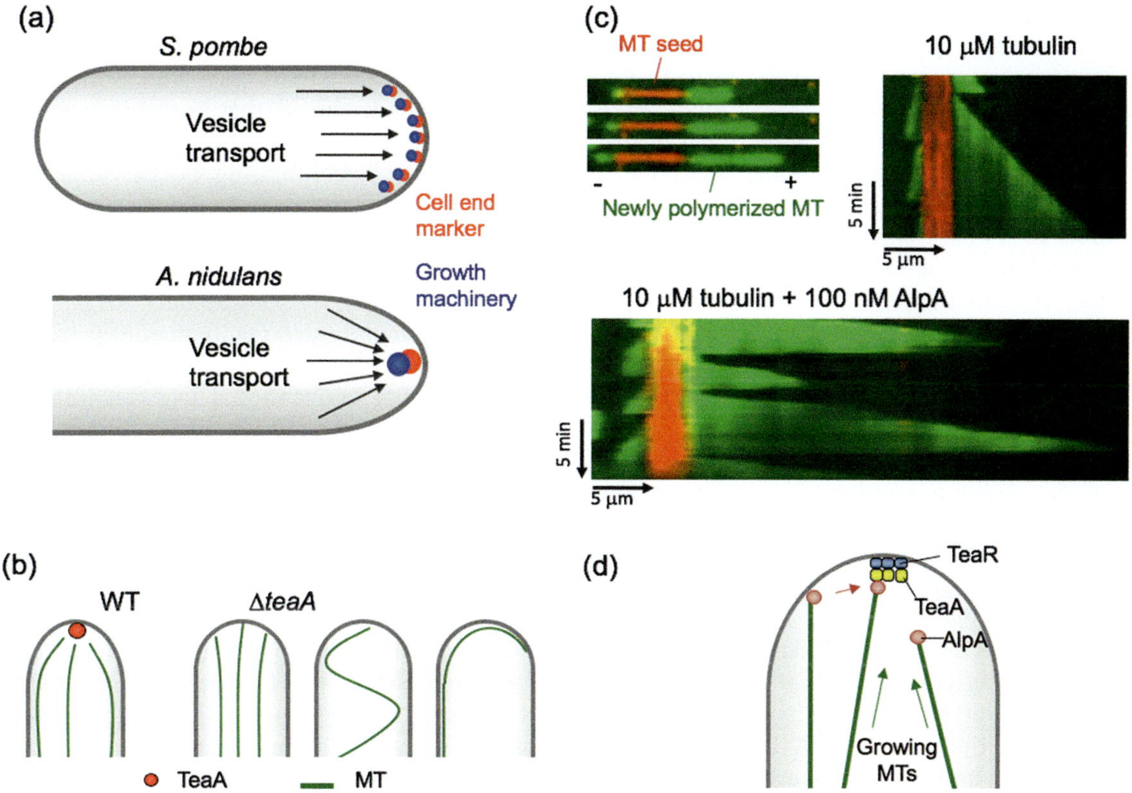

Fig. 5 Cell-end markers for polarity focusing. (**a**) Comparison of the localization of cell-end markers and the growth machinery in *S. pombe* and *A. nidulans*. (**b**) Behavior of MTs at hyphal tips in *A. nidulans* wild type and Δ*teaA* strains. (**c**) In vitro MT polymerization assay. Images of a seed MT (red) with a dynamic MT lattice growing from the plus end (green). Kymographs of MTs in the absence of AlpA (right) and presence of 100 nM AlpA (lower). (**d**) Scheme of the interaction between TeaA at the hyphal tip cortex and AlpA at MT plus ends

mechanism is conserved in *S. pombe* as well (Snaith and Sawin 2003; Takeshita et al. 2008; Bicho et al. 2010). MTs in *A. nidulans* elongate toward the tips and tend to converge in the apical region (Konzack et al. 2005), which is not observed in *S. pombe*. The central position of TeaA at the tip correlated with the convergence of the MT plus ends to a single point. In the absence of TeaA, MTs often contacted the membrane off the center of the apex (Fig. 5b) (Takeshita et al. 2008, 2013).

A recent study showed that a functional connection between TeaA and the **MT polymerase AlpA** is required for proper regulation of MT growth at hyphal tips (Takeshita et al. 2013). AlpA is a member of the XMAP215/Dis1 family whose conserved TOG domains, which contain multiple HEAT repeats, are known to bind tubulin from yeast to human (Al-Bassam et al. 2007). XMAP215 from *Xenopus laevis* catalyzes the addition of tubulin dimers to the growing plus ends (Brouhard et al. 2008; Al-Bassam and Chang 2011). *A. nidulans* AlpA decorates MT filaments and accumulates at MT plus ends (Enke et al. 2007). Deletion of *alpA* resulted in a drastic reduction of the MT array and dynamics. MT in vitro polymerization assays with purified tubulin from porcine brains and recombinant AlpA have revealed the activity of AlpA as a MT polymerase (Fig. 5c) (Takeshita et al. 2013). The MT growth speed in vitro was comparable with that of XMAP215 of *X. laevis* and approximately fourfold higher than that of Alp14, the orthologue in *S. pombe* (Brouhard et al. 2008; Al-Bassam et al. 2012). The rate of MT polymerization in vivo in *A. nidulans* leading hyphae is approximately threefold higher than in *S. pombe*, consistent with the ratio in vitro (Drummond and Cross 2000; Efimov et al. 2006). However, AlpA-dependent MT growth speed in vitro was approximately only half of the one determined in vivo (6 μm/min compared to 13 ± 3 μm/min). Therefore, other microtubule plus-end-tracking proteins are likely to enhance the AlpA activity for MT growth in vivo.

As a difference to *S. pombe*, *A. nidulans* TeaA is involved in the convergence of MT plus ends at the tip apex, suggesting specific interactions of the MT plus end with the cortex. One possibility is an interaction between TeaA and AlpA (Takeshita et al. 2013). MT polymerization assays in vitro showed that TeaA increased the catastrophe frequency of MTs in the presence of AlpA, and TeaA reduced the in vitro AlpA activity significantly. From these results it was concluded that AlpA promotes MT growth at MT plus ends until MTs reach the hyphal tip, where TeaA blocks the AlpA activity and induces MT catastrophe (Fig. 5d). The interdependence of TeaA and MTs could act as a positive feedback loop to concentrate TeaA at the apex resulting in well-focused vesicle secretion for the organization of the Spitzenkörper (Bartnicki-Garcia et al. 1995; Riquelme et al. 2014).

X. Transient Polarity Marker Assembly

The membrane-associated cell-end marker TeaR is highly dynamic at growing hyphal tips (Ishitsuka et al. 2015). A MT grows toward the hyphal tip, pauses in close contact with the apical membrane, and then undergoes a catastrophe event resulting in retraction. TeaR accumulates at the hyphal tip membrane and then decreases immediately after a MT plus-end touches the tip membrane and starts to shrink. Colocalization studies indicate that TeaR clusters represent zones of exocytosis at the apical membrane. In general, membrane-binding polarity markers are delivered to the plasma membrane by vesicle transport and exocytosis. After exocytosis, vesicles fuse with the plasma membrane, leading to its extension. Simulation analysis had predicted that membrane insertion by active exocytosis would dilute and/or disperse membrane-binding polarity markers during fast hyphal growth (Savage et al. 2012). There was no clear answer to the question of how cell polarity is maintained during incessant vesicle exocytosis, especially for rapidly growing systems such as filamentous fungi. One of the problems to capture the complex process is that conventional live cell imaging methods lack the resolution.

The localization of TeaR cluster at hyphal tips has been analyzed by super-resolution microscopy PALM (photoactivation localization microscopy) (Ishitsuka et al. 2015). The resolution of conventional light microscopy techniques is limited to about 250 nm due to light diffraction. Super-resolution microscopy techniques, such as PALM, have overcome the diffraction limit, resulting in lateral image resolution as high as 20 nm and providing a powerful tool for the investigation of protein localization in high detail (Sahl and Moerner 2013). The PALM imaging analysis revealed that TeaR transiently assembles (approximately 120 nm, 20 TeaR proteins per cluster on average) at the hyphal tip membrane and disperses along the membrane after exocytosis, which inserts a new membrane that results in local membrane extension (Fig. 6a) (Ishitsuka et al. 2015).

These findings gave rise to a "transient polarity assembly model" to explain how fungal tip cells extend through repeated cycles of TeaR assembly/disassembly, actin polymerization, and exocytosis rather than by constant elongation (Fig. 6b) (Takeshita 2016). The findings of colocalization studies support the notion that TeaR clusters represent zones of exocytosis and are prerequisite for apical membrane extension. In this model, the interaction between the MT plus end, where TeaA is located, and TeaR at the apical membrane initiates the recruitment of other polarity markers, resulting in the assembly of TeaR polarity site. The accumulated cell-end markers induce actin polymerization followed by active exocytosis.

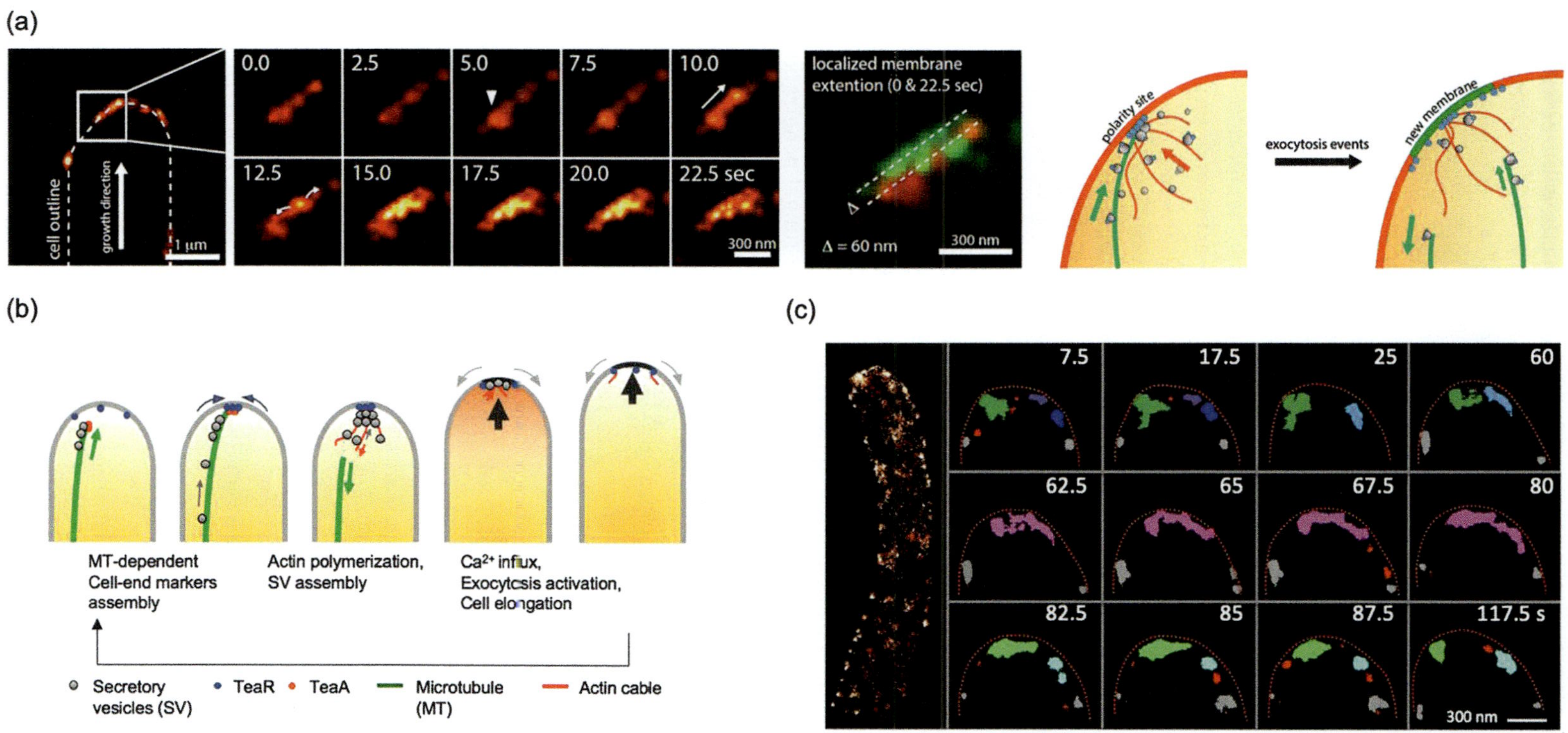

Fig. 6 Oscillatory fungal cell growth. (**a**) Time-lapse PALM of TeaR shows the appearance of a new cluster ($t = 5$ s), a translational movement ($t = 10$ s), and a spreading of the signal along the plasma membrane. Local membrane extension ($t = 0$ in red, $t = 22.5$ s in green). (**b**) Scheme of transient polarity model and temporally controlled actin polymerization and exocytosis coordinated by pulsed Ca^{2+} influx. (**c**) Time-lapse cluster analysis of PALM images of chitin synthase ChsB. (Left) Localization image of a hypha with mEosFPthermo-ChsB clusters (500 frames). (Right) Sequence of ChsB cluster images (clusters in different colors) rendered from images reconstructed by moving window binning (cluster images of 2.5 s time intervals for a total period of 125 s)

Newly synthesized TeaR is delivered to the tip membrane on secretory vesicles through MT and actin cables. The plasma membrane extends locally at the site of vesicle fusion, and, subsequently, the TeaR polarity site is dispersed or displaced along the membrane. Once the polarity site is disassembled, however, next MT comes to the tip and gathers TeaR floating in the membrane through the interaction with TeaA at the MT plus end and the cycle starts over.

In line with this model, recent work on *N. crassa* has identified bursts of exocytotic events at various sites within the apical membrane rather than a persistent exocytosis site (Riquelme et al. 2014). Mathematical simulation analysis confirmed the validity of the transient polarity assembly model (Ishitsuka et al. 2015). The simultaneous visualization of actin cables and MTs suggests temporally and spatially coordinated polymerization and depolymerization between the two cytoskeletons (Bergs et al. 2016). Interaction between MigA, a MT plus-end localizing protein, and a class V myosin suggests that an active mechanism captures MTs and pulls the ends along actin filaments (Manck et al. 2015). These results also support the model.

XI. Oscillatory Fungal Cell Growth

Many dynamic cellular processes that appear continuous are driven by underlying mechanisms that oscillate with distinct periods. For example, eukaryotic cells do not grow continuously but rather by pulsed extension of the periphery. Stepwise cell extension at the hyphal tips of several filamentous fungi was discovered 20 years ago (Lopez-Franco et al. 1994), but only a few molecular details of the mechanism have been clarified since then. The "transient polarity assembly model" also implies the stepwise cell extension. Indeed, the time-lapse PALM imaging revealed that the cell extension rate of hyphae is not constant but varies in an oscillatory manner (Fig. 6c) (Zhou et al. 2018).

In addition, a recent study has provided evidence for molecular mechanism of oscillatory fungal cell growth. Live cell imaging analyses revealed oscillations of actin assembly and exocytosis in growing hyphal tips (Takeshita et al. 2017). Intracellular Ca^{2+} levels are known to regulate actin assembly and vesicle fusion (Janmey 1994; Schneggenburger and Neher 2005). Ca^{2+} levels also pulsed at the hyphal tips (Kim et al. 2012). The **fluorescence-based Ca^{2+} biosensor R-GECO** varies the emission according to Ca^{2+} concentrations and has enabled the visualization of pulsatile Ca^{2+} concentrations in growing hyphal tips of *A. nidulans* (Takeshita et al. 2017). Intracellular Ca^{2+} levels pulse at ~30-second intervals. These positively and temporally correlate with amounts of F-actin and secretory vesicles at hyphal tips. Orthologues of Ca^{2+} channels at the plasma membrane in *A. nidulans* are required for proper tip growth and the oscillation of F-actin, secretory vesicles, and Ca^{2+} level pulses (Wang et al. 2012; Takeshita et al. 2017). These results suggest that pulsed Ca^{2+} influx coordinates the temporally controlled actin polymerization and exocytosis that drive stepwise cell extension (Fig. 6b). The oscillatory fungal growth could be important for dynamic responses to external and internal signals in chemotropism, cell-cell fusion, microbial interaction, and the fungal penetration of plant and animal cells (Takeshita 2018).

XII. Conclusion

The establishment and maintenance of cell polarity in fungi—as in higher eukaryotes—require the interplay between the actin and MT cytoskeletons and landmark proteins at the cortex (Siegrist and Doe 2007; Li and Gundersen 2008). This rather complex arrangement of components may be necessary to guarantee robust polar growth. Coordinated oscillations of actin polymerization, exocytosis, and Ca^{2+} levels associated with cell growth seem to be a conserved phenomenon shared among various organisms, including fungi (Das et al. 2012), mammalian cells (Wollman and Meyer 2012), and plant root hairs (Monshausen et al. 2008) and pollen tubes (Kroeger and Geitmann 2012). The oscillatory cell growth allows cells to respond more quickly and often to internal and external cues such as chemical or mechanical environmental signals.

References

Abenza JF, Pantazopoulou A, Rodriguez JM, Galindo A, Penalva MA (2009) Long-distance movement of *Aspergillus nidulans* early endosomes on microtubule tracks. Traffic 10:57–75

Adams AE, Johnson DI, Longnecker RM, Sloat BF, Pringle JR (1990) CDC42 and CDC43, two additional genes involved in budding and the establishment of cell polarity in the yeast *Saccharomyces cerevisiae*. J Cell Biol 111: 131–142

Al-Bassam J, Chang F (2011) Regulation of microtubule dynamics by TOG-domain proteins XMAP215/Dis1 and CLASP. Trends Cell Biol 21:604–614

Al-Bassam J, Larsen NA, Hyman AA, Harrison SC (2007) Crystal structure of a TOG domain: conserved features of XMAP215/Dis1-family TOG domains and implications for tubulin binding. Structure 15:355–362

Al-Bassam J, Kim H, Flor-Parra I, Lal N, Velji H, Chang F (2012) Fission yeast Alp14 is a dose-dependent plus end-tracking microtubule polymerase. Mol Biol Cell 23:2878–2890

Alvarez-Tabares I, Perez-Martin J (2010) Septins from the phytopathogenic fungus *Ustilago maydis* are required for proper morphogenesis but dispensable for virulence. PLoS One 5:e12933

Amberg DC (1998) Three-dimensional imaging of the yeast actin cytoskeleton through the budding cell cycle. Mol Biol Cell 9:3259–3262

Araujo-Bazan L, Penalva MA, Espeso EA (2008) Preferential localization of the endocytic internalization machinery to hyphal tips underlies polarization of the actin cytoskeleton in *Aspergillus nidulans*. Mol Microbiol 67:891–905

Asakura T, Sasaki T, Nagano F, Satoh A, Obaishi H, Nishioka H, Imamura H, Hotta K, Tanaka K, Nakanishi H, Takai Y (1998) Isolation and characterization of a novel actin filament-binding protein from *Saccharomyces cerevisiae*. Oncogene 16:121–130

Bartnicki-García S, Hergert F, Gierz G (1989) Computer simulation of fungal morphogenesis and the mathematical basis for hyphal tip growth. Protoplasma 153:46–57

Bartnicki-Garcia S, Bartnicki DD, Gierz G, Lopez-Franco R, Bracker CE (1995) Evidence that Spitzenkörper behavior determines the shape of a fungal hypha: a test of the hyphoid model. Exp Mycol 19:153–159

Baumann S, Pohlmann T, Jungbluth M, Brachmann A, Feldbrugge M (2012) Kinesin-3 and dynein mediate microtubule-dependent co-transport of mRNPs and endosomes. J Cell Sci 125:2740–2752

Baumann S, Konig J, Koepke J, Feldbrugge M (2014) Endosomal transport of septin mRNA and protein indicates local translation on endosomes and is required for correct septin filamentation. EMBO Rep 15:94–102

Berepiki A, Read ND (2013) Septins are important for cell polarity, septation and asexual spore formation in *Neurospora crassa* and show different patterns of localisation at germ tube tips. PLoS One 8: e63843

Berepiki A, Lichius A, Shoji JY, Tilsner J, Read ND (2010) F-actin dynamics in *Neurospora crassa*. Eukaryot Cell 9:547–557

Berepiki A, Lichius A, Read ND (2011) Actin organization and dynamics in filamentous fungi. Nat Rev Microbiol 9:876–887

Bergs A, Ishitsuka Y, Evangelinos M, Nienhaus GU, Takeshita N (2016) Dynamics of actin cables in polarized growth of the filamentous fungus *Aspergillus nidulans*. Front Microbiol 7:682

Bernards A, Settleman J (2004) GAP control: regulating the regulators of small GTPases. Trends Cell Biol 14:377–385

Bertin A, McMurray MA, Pierson J, Thai L, McDonald KL, Zehr EA, Garcia G 3rd, Peters P, Thorner J, Nogales E (2012) Three-dimensional ultrastructure of the septin filament network in *Saccharomyces cerevisiae*. Mol Biol Cell 23:423–432

Bicho CC, Kelly DA, Snaith HA, Goryachev AB, Sawin KE (2010) A catalytic role for Mod5 in the formation of the Tea1 cell polarity landmark. Curr Biol 20:1752–1757

Bishop AL, Hall A (2000) Rho GTPases and their effector proteins. Biochem J 348(Pt 2):241–255

Boyce KJ, Chang H, D'Souza CA, Kronstad JW (2005) An *Ustilago maydis* septin is required for filamentous growth in culture and for full symptom development on maize. Eukaryot Cell 4:2044–2056

Brent Heath I, Bonham M, Akram A, Gupta GD (2003) The interrelationships of actin and hyphal tip growth in the ascomycete *Geotrichum candidum*. Fungal Genet Biol 38:85–97

Bridges AA, Gladfelter AS (2014) Fungal pathogens are platforms for discovering novel and conserved septin properties. Curr Opin Microbiol 20:42–48

Brouhard GJ, Stear JH, Noetzel TL, Al-Bassam J, Kinoshita K, Harrison SC, Howard J, Hyman AA (2008) XMAP215 is a processive microtubule polymerase. Cell 132:79–88

Browning H, Hayles J, Mata J, Aveline L, Nurse P, McIntosh JR (2000) Tea2p is a kinesin-like protein required to generate polarized growth in fission yeast. J Cell Biol 151:15–28

Browning H, Hackney DD, Nurse P (2003) Targeted movement of cell end factors in fission yeast. Nat Cell Biol 5:812–818

Chang F, Peter M (2003) Yeasts make their mark. Nat Cell Biol 5:294–299

Dagdas YF, Yoshino K, Dagdas G, Ryder LS, Bielska E, Steinberg G, Talbot NJ (2012) Septin-mediated plant cell invasion by the rice blast fungus, *Magnaporthe oryzae*. Science 336:1590–1595

Das M, Drake T, Wiley DJ, Buchwald P, Vavylonis D, Verde F (2012) Oscillatory dynamics of Cdc42

GTPase in the control of polarized growth. Science 337:239–243
Delgado-Alvarez DL, Callejas-Negrete OA, Gomez N, Freitag M, Roberson RW, Smith LG, Mourino-Perez RR (2010) Visualization of F-actin localization and dynamics with live cell markers in *Neurospora crassa*. Fungal Genet Biol 47:573–586
Delgado-Alvarez DL, Bartnicki-Garcia S, Seiler S, Mourino-Perez RR (2014) Septum development in *Neurospora crassa*: the septal actomyosin tangle. PLoS One 9:e96744
DeMay BS, Meseroll RA, Occhipinti P, Gladfelter AS (2009) Regulation of distinct septin rings in a single cell by Elm1p and Gin4p kinases. Mol Biol Cell 20:2311–2326
Dodgson J, Chessel A, Yamamoto M, Vaggi F, Cox S, Rosten E, Albrecht D, Geymonat M, Csikasz-Nagy A, Sato M, Carazo-Salas RE (2013) Spatial segregation of polarity factors into distinct cortical clusters is required for cell polarity control. Nat Commun 4:1834
Dominguez R, Holmes KC (2011) Actin structure and function. Annu Rev Biophys 40:169–186
Drummond DR, Cross RA (2000) Dynamics of interphase microtubules in *Schizosaccharomyces pombe*. Curr Biol 10:766–775
Efimov VP, Zhang J, Xiang X (2006) CLIP-170 homologue and NUDE play overlapping roles in NUDF localization in *Aspergillus nidulans*. Mol Biol Cell 17:2021–2034
Egan MJ, McClintock MA, Reck-Peterson SL (2012a) Microtubule-based transport in filamentous fungi. Curr Opin Microbiol 15:637–645
Egan MJ, Tan K, Reck-Peterson SL (2012b) Lis1 is an initiation factor for dynein-driven organelle transport. J Cell Biol 197:971–982
Enke C, Zekert N, Veith D, Schaaf C, Konzack S, Fischer R (2007) *Aspergillus nidulans* Dis1/XMAP215 protein AlpA localizes to spindle pole bodies and microtubule plus ends and contributes to growth directionality. Eukaryot Cell 6:555–562
Evangelista M, Pruyne D, Amberg DC, Boone C, Bretscher A (2002) Formins direct Arp2/3-independent actin filament assembly to polarize cell growth in yeast. Nat Cell Biol 4:260–269
Fajardo-Somera RA, Jöhnk B, Bayram O, Valerius O, Braus GH, Riquelme M (2015) Dissecting the function of the different chitin synthases in vegetative growth and sexual development in *Neurospora crassa*. Fungal Genet Biol 75:30–45
Feierbach B, Verde F, Chang F (2004) Regulation of a formin complex by the microtubule plus end protein tea1p. J Cell Biol 165:697–707
Fischer R, Zekert N, Takeshita N (2008) Polarized growth in fungi–interplay between the cytoskeleton, positional markers and membrane domains. Mol Microbiol 68:813–826
Garcia-Vidal C, Viasus D, Carratala J (2013) Pathogenesis of invasive fungal infections. Curr Opin Infect Dis 26:270–276
Gladfelter AS (2010) Guides to the final frontier of the cytoskeleton: septins in filamentous fungi. Curr Opin Microbiol 13:720–726
Goehring NW, Grill SW (2013) Cell polarity: mechanochemical patterning. Trends Cell Biol 23:72–80
Grove SN, Bracker CE (1970) Protoplasmic organization of hyphal tips among fungi: vesicles and Spitzenkörper. J Bacteriol 104:989–1009
Gunning PW, Schevzov G, Kee AJ, Hardeman EC (2005) Tropomyosin isoforms: divining rods for actin cytoskeleton function. Trends Cell Biol 15:333–341
Haag C, Steuten B, Feldbrügge M (2015) Membrane-coupled mRNA trafficking in fungi. Annu Rev Microbiol 69:265–281
Harris SD (2001) Septum formation in *Aspergillus nidulans*. Curr Opin Microbiol 4:736–739
Harris SD (2006) Cell polarity in filamentous fungi: shaping the mold. Int Rev Cytol 251:41–77
Harris SD (2008) Branching of fungal hyphae: regulation, mechanisms and comparison with other branching systems. Mycologia 100:823–832
Harris SD, Momany M (2004) Polarity in filamentous fungi: moving beyond the yeast paradigm. Fungal Genet Biol 41:391–400
Harris SD, Hofmann AF, Tedford HW, Lee MP (1999) Identification and characterization of genes required for hyphal morphogenesis in the filamentous fungus *Aspergillus nidulans*. Genetics 151:1015–1025
Harris SD, Read ND, Roberson RW, Shaw B, Seiler S, Plamann M, Momany M (2005) Polarisome meets spitzenkörper: microscopy, genetics, and genomics converge. Eukaryot Cell 4:225–229
Helfer H, Gladfelter AS (2006) AgSwe1p regulates mitosis in response to morphogenesis and nutrients in multinucleated *Ashbya gossypii* cells. Mol Biol Cell 17:4494–4512
Hernandez-Rodriguez Y, Masuo S, Johnson D, Orlando R, Smith A, Couto-Rodriguez M, Momany M (2014) Distinct septin heteropolymers co-exist during multicellular development in the filamentous fungus *Aspergillus nidulans*. PLoS One 9: e92819
Hervas-Aguilar A, Penalva MA (2010) Endocytic machinery protein SlaB is dispensable for polarity establishment but necessary for polarity maintenance in hyphal tip cells of *Aspergillus nidulans*. Eukaryot Cell 9:1504–1518
Higashitsuji Y, Herrero S, Takeshita N, Fischer R (2009) The cell end marker protein TeaC is involved in growth directionality and septation in *Aspergillus nidulans*. Eukaryot Cell 8:957–967
Higuchi Y, Ashwin P, Roger Y, Steinberg G (2014) Early endosome motility spatially organizes polysome distribution. J Cell Biol 204:343–357
Hill TW, Jackson-Hayes L, Wang X, Hoge BL (2015) A mutation in the converter subdomain of *Aspergillus nidulans* MyoB blocks constriction of the

actomyosin ring in cytokinesis. Fungal Genet Biol 75:72–83

Horio T, Oakley BR (2005) The role of microtubules in rapid hyphal tip growth of *Aspergillus nidulans*. Mol Biol Cell 16:918–926

Ishitsuka Y, Savage N, Li Y, Bergs A, Grün N, Kohler D, Donnelly R, Nienhaus GU, Fischer R, Takeshita N (2015) Super-resolution microscopy reveals a dynamic picture of cell polarity maintenance during directional growth. Sci Adv 1:e1500947

Jaffe AB, Hall A (2005) Rho GTPases: biochemistry and biology. Annu Rev Cell Dev Biol 21:247–269

Janmey PA (1994) Phosphoinositides and calcium as regulators of cellular actin assembly and disassembly. Annu Rev Physiol 56:169–191

Johnson JM, Jin M, Lew DJ (2011) Symmetry breaking and the establishment of cell polarity in budding yeast. Curr Opin Genet Dev 21:740–746

Kaksonen M, Toret CP, Drubin DG (2006) Harnessing actin dynamics for clathrin-mediated endocytosis. Nat Rev Mol Cell Biol 7:404–414

Kim HS, Czymmek KJ, Patel A, Modla S, Nohe A, Duncan R, Gilroy S, Kang S (2012) Expression of the Cameleon calcium biosensor in fungi reveals distinct Ca(2+) signatures associated with polarized growth, development, and pathogenesis. Fungal Genet Biol 49:589–601

Kohli M, Galati V, Boudier K, Roberson RW, Philippsen P (2008) Growth-speed-correlated localization of exocyst and polarisome components in growth zones of *Ashbya gossypii* hyphal tips. J Cell Sci 121:3878–3889

Konig J, Baumann S, Koepke J, Pohlmann T, Zarnack K, Feldbrugge M (2009) The fungal RNA-binding protein Rrm4 mediates long-distance transport of ubi1 and rho3 mRNAs. EMBO J 28:1855–1866

Konzack S, Rischitor PE, Enke C, Fischer R (2005) The role of the kinesin motor KipA in microtubule organization and polarized growth of *Aspergillus nidulans*. Mol Biol Cell 16:497–506

Kozubowski L, Heitman J (2010) Septins enforce morphogenetic events during sexual reproduction and contribute to virulence of *Cryptococcus neoformans*. Mol Microbiol 75:658–675

Kroeger J, Geitmann A (2012) The pollen tube paradigm revisited. Curr Opin Plant Biol 15:618–624

Lee SC, Schmidtke SN, Dangott LJ, Shaw BD (2008) *Aspergillus nidulans* ArfB plays a role in endocytosis and polarized growth. Eukaryot Cell 7:1278–1288

Lehmler C, Steinberg G, Snetselaar KM, Schliwa M, Kahmann R, Bolker M (1997) Identification of a motor protein required for filamentous growth in *Ustilago maydis*. EMBO J 16:3464–3473

Li R, Gundersen GG (2008) Beyond polymer polarity: how the cytoskeleton builds a polarized cell. Nat Rev Mol Cell Biol 9:860–873

Lichius A, Goryachev AB, Fricker MD, Obara B, Castro-Longoria E, Read ND (2014) CDC-42 and RAC-1 regulate opposite chemotropisms in *Neurospora crassa*. J Cell Sci 127:1953–1965

Lindsey R, Momany M (2006) Septin localization across kingdoms: three themes with variations. Curr Opin Microbiol 9:559–565

Lindsey R, Ha Y, Momany M (2010) A septin from the filamentous fungus *A. nidulans* induces atypical pseudohyphae in the budding yeast *S. cerevisiae*. PLoS One 5:e9858

Lopez-Franco R, Bartnicki-Garcia S, Bracker CE (1994) Pulsed growth of fungal hyphal tips. Proc Natl Acad Sci 91:12228–12232

Mahlert M, Leveleki L, Hlubek A, Sandrock B, Bolker M (2006) Rac1 and Cdc42 regulate hyphal growth and cytokinesis in the dimorphic fungus *Ustilago maydis*. Mol Microbiol 59:567–578

Manck R, Ishitsuka Y, Herrero S, Takeshita N, Nienhaus GU, Fischer R (2015) Genetic evidence for a microtubule-capture mechanism during polarised growth of *Aspergillus nidulans*. J Cell Sci 128: 3569–3582

Markina-Inarrairaegui A, Pantazopoulou A, Espeso EA, Penalva MA (2013) The *Aspergillus nidulans* peripheral ER: disorganization by ER stress and persistence during mitosis. PLoS One 8:e67154

Martin SG, Chang F (2003) Cell polarity: a new mod(e) of anchoring. Curr Biol 13:R711–R713

Mata J, Nurse P (1997) tea1 and the microtubular cytoskeleton are important for generating global spatial order within the fission yeast cell. Cell 89:939–949

Meyer SL, Kaminskyj SG, Heath IB (1988) Nuclear migration in a nud mutant of *Aspergillus nidulans* is inhibited in the presence of a quantitatively normal population of cytoplasmic microtubules. J Cell Biol 106:773–778

Michelot A, Drubin DG (2011) Building distinct actin filament networks in a common cytoplasm. Curr Biol 21:R560–R569

Miller PJ, Johnson DI (1994) Cdc42p GTPase is involved in controlling polarized cell growth in *Schizosaccharomyces pombe*. Mol Cell Biol 14: 1075–1083

Momany M, Hamer JE (1997) Relationship of actin, microtubules, and crosswall synthesis during septation in *Aspergillus nidulans*. Cell Motil Cytoskeleton 38:373–384

Momany M, Westfall PJ, Abramowsky G (1999) *Aspergillus nidulans* swo mutants show defects in polarity establishment, polarity maintenance and hyphal morphogenesis. Genetics 151:557–567

Monshausen GB, Messerli MA, Gilroy S (2008) Imaging of the Yellow Cameleon 3.6 indicator reveals that elevations in cytosolic Ca2+ follow oscillating increases in growth in root hairs of *Arabidopsis*. Plant Physiol 147:1690–1698

Montegi F, Arai R, Mabuchi I (2001) Identification of two type V myosins in fission yeast, one of which functions in polarized cell growth and moves rapidly in the cell. Mol Biol Cell 12:1367–1380

Morris NR, Enos AP (1992) Mitotic gold in a mold: *Aspergillus* genetics and the biology of mitosis. Trends Genet 8:32–37
Morris NR, Lai MH, Oakley CE (1979) Identification of a gene for alpha-tubulin in *Aspergillus nidulans*. Cell 16:437–442
Mostowy S, Cossart P (2012) Septins: the fourth component of the cytoskeleton. Nat Rev Mol Cell Biol 13:183–194
Mulvihill DP, Edwards SR, Hyams JS (2006) A critical role for the type V myosin, Myo52, in septum deposition and cell fission during cytokinesis in *Schizosaccharomyces pombe*. Cell Motil Cytoskeleton 63:149–161
Oakley BR (2004) Tubulins in *Aspergillus nidulans*. Fungal Genet Biol 41:420–427
Oakley BR, Morris NR (1980) Nuclear movement is beta-tubulin-dependent in *Aspergillus nidulans*. Cell 19:255–262
Oakley CE, Oakley BR (1989) Identification of gamma-tubulin, a new member of the tubulin superfamily encoded by mipA gene of *Aspergillus nidulans*. Nature 338:662–664
Oakley BR, Oakley CE, Yoon Y, Jung MK (1990) Gamma-tubulin is a component of the spindle pole body that is essential for microtubule function in *Aspergillus nidulans*. Cell 61:1289–1301
Oh Y, Bi E (2011) Septin structure and function in yeast and beyond. Trends Cell Biol 21:141–148
Orr E, Rosenberger RF (1976) Initial characterization of *Aspergillus nidulans* mutants blocked in the nuclear replication cycle. J Bacteriol 126:895–902
Osmani SA, Mirabito PM (2004) The early impact of genetics on our understanding of cell cycle regulation in *Aspergillus nidulans*. Fungal Genet Biol 41:401–410
Pantazopoulou A, Pinar M, Xiang X, Penalva MA (2014) Maturation of late Golgi cisternae into RabE (RAB11) exocytic post-Golgi carriers visualized *in vivo*. Mol Biol Cell 25:2428–2443
Pearson CL, Xu K, Sharpless KE, Harris SD (2004) MesA, a novel fungal protein required for the stabilization of polarity axes in *Aspergillus nidulans*. Mol Biol Cell 15:3658–3672
Pelham RJ Jr, Chang F (2001) Role of actin polymerization and actin cables in actin-patch movement in *Schizosaccharomyces pombe*. Nat Cell Biol 3:235–244
Penalva MA (2010) Endocytosis in filamentous fungi: Cinderella gets her reward. Curr Opin Microbiol 13:684–692
Pinar M, Pantazopoulou A, Arst HN Jr, Penalva MA (2013) Acute inactivation of the *Aspergillus nidulans* Golgi membrane fusion machinery: correlation of apical extension arrest and tip swelling with cisternal disorganization. Mol Microbiol 89:228–248
Punt PJ, Levasseur A, Visser H, Wery J, Record E (2011) Fungal protein production: design and production of chimeric proteins. Annu Rev Microbiol 65:57–69
Requena N, Alberti-Segui C, Winzenburg E, Horn C, Schliwa M, Philippsen P, Liese R, Fischer R (2001) Genetic evidence for a microtubule-destabilizing effect of conventional kinesin and analysis of its consequences for the control of nuclear distribution in *Aspergillus nidulans*. Mol Microbiol 42:121–132
Riedl J, Crevenna AH, Kessenbrock K, Yu JH, Neukirchen D, Bista M, Bradke F, Jenne D, Holak TA, Werb Z, Sixt M, Wedlich-Soldner R (2008) Lifeact: a versatile marker to visualize F-actin. Nat Methods 5:605–607
Riquelme M, Sanchez-Leon E (2014) The Spitzenkörper: a choreographer of fungal growth and morphogenesis. Curr Opin Microbiol 20:27–33
Riquelme M, Reynaga-Pena CG, Gierz G, Bartnicki-Garcia S (1998) What determines growth direction in fungal hyphae? Fungal Genet Biol 24:101–109
Riquelme M, Gierz G, Bartnicki-García S (2000) Dynein and dynactin deficiencies affect the formation and function of the Spitzenkörper and distort hyphal morphogenesis of Neurospora crassa. Microbiology 146:1743–1752
Riquelme M, Bartnicki-García S, González-Prieto JM, Sánchez-León E, Verdín-Ramos JA, Beltrán-Aguilar A, Freitag M (2007) Spitzenkörper localization and intracellular traffic of green fluorescent protein-labeled CHS-3 and CHS-6 chitin synthases in living hyphae of *Neurospora crassa*. Eukaryot Cell 6:1853–1864
Riquelme M, Yarden O, Bartnicki-Garcia S, Bowman B, Castro-Longoria E, Free SJ, Fleissner A, Freitag M, Lew RR, Mourino-Perez R, Plamann M, Rasmussen C, Richthammer C, Roberson RW, Sanchez-Leon E, Seiler S, Watters MK (2011) Architecture and development of the *Neurospora crassa* hypha – a model cell for polarized growth. Fungal Biol 115:446–474
Riquelme M, Bredeweg EL, Callejas-Negrete O, Roberson RW, Ludwig S, Beltran-Aguilar A, Seiler S, Novick P, Freitag M (2014) The *Neurospora crassa* exocyst complex tethers Spitzenkörper vesicles to the apical plasma membrane during polarized growth. Mol Biol Cell 25:1312–1326
Riquelme M, Aguirre J, Bartnicki-García S, Braus GH, Feldbrügge M, Fleig U, Hansberg W, Herrera-Estrella A, Kämper J, Kück U, Mouriño-Pérez RR, Takeshita N, Fischer R (2018) Fungal morphogenesis, from the polarized growth of hyphae to complex reproduction and infection structures. Microbiol Mol Biol Rev 82(2)
Sahl SJ, Moerner WE (2013) Super-resolution fluorescence imaging with single molecules. Curr Opin Struct Biol 23:778–787
Sanchez-Leon E, Verdin J, Freitag M, Roberson RW, Bartnicki-Garcia S, Riquelme M (2011) Traffic of chitin synthase 1 (CHS-1) to the Spitzenkörper and developing septa in hyphae of *Neurospora crassa*: actin dependence and evidence of distinct microvesicle populations. Eukaryot Cell 10:683–695
Saunders DG, Dagdas YF, Talbot NJ (2010) Spatial uncoupling of mitosis and cytokinesis during appressorium-mediated plant infection by the

rice blast fungus *Magnaporthe oryzae*. Plant Cell 22:2417–2428

Savage NS, Layton AT, Lew DJ (2012) Mechanistic mathematical model of polarity in yeast. Mol Biol Cell 23:1998–2013

Sawin KE, Tran PT (2006) Cytoplasmic microtubule organization in fission yeast. Yeast 23:1001–1014

Schmidt A, Hall A (2002) Guanine nucleotide exchange factors for Rho GTPases: turning on the switch. Genes Dev 16:1587–1609

Schneggenburger R, Neher E (2005) Presynaptic calcium and control of vesicle fusion. Curr Opin Neurobiol 15:266–274

Schoch CL, Aist JR, Yoder OC, Gillian Turgeon B (2003) A complete inventory of fungal kinesins in representative filamentous ascomycetes. Fungal Genet Biol 39:1–15

Schultzhaus Z, Yan H, Shaw BD (2015) *Aspergillus nidulans* flippase DnfA is cargo of the endocytic collar and plays complementary roles in growth and phosphatidylserine asymmetry with another flippase, DnfB. Mol Microbiol 97:18–32

Schuster M, Treitschke S, Kilaru S, Molloy J, Harmer NJ, Steinberg G (2012) Myosin-5, kinesin-1 and myosin-17 cooperate in secretion of fungal chitin synthase. EMBO J 31:214–227

Seidel C, Zekert N, Fischer R (2012) The *Aspergillus nidulans* kinesin-3 tail is necessary and sufficient to recognize modified microtubules. PLoS One 7: e30976

Seiler S, Nargang FE, Steinberg G, Schliwa M (1997) Kinesin is essential for cell morphogenesis and polarized secretion in *Neurospora crassa*. EMBO J 16:3025–3034

Seiler S, Plamann M, Schliwa M (1999) Kinesin and dynein mutants provide novel insights into the roles of vesicle traffic during cell morphogenesis in *Neurospora*. Curr Biol 9:779–785

Shaw BD, Momany C, Momany M (2002) *Aspergillus nidulans* swoF encodes an N-myristoyl transferase. Eukaryot Cell 1:241–248

Shaw BD, Chung DW, Wang CL, Quintanilla LA, Upadhyay S (2011) A role for endocytic recycling in hyphal growth. Fungal Biol 115:541–546

Siegrist SE, Doe CQ (2007) Microtubule-induced cortical cell polarity. Genes Dev 21:483–496

Sinha I, Wang YM, Philp R, Li CR, Yap WH, Wang Y (2007) Cyclin-dependent kinases control septin phosphorylation in *Candida albicans* hyphal development. Dev Cell 13:421–432

Snaith HA, Sawin KE (2003) Fission yeast mod5p regulates polarized growth through anchoring of tea1p at cell tips. Nature 423:647–651

Steinberg G (2011) Motors in fungal morphogenesis: cooperation versus competition. Curr Opin Microbiol 14:660–667

Steinberg G (2014) Endocytosis and early endosome motility in filamentous fungi. Curr Opin Microbiol 20:10–18

Sudbery PE (2001) The germ tubes of *Candida albicans* hyphae and pseudohyphae show different patterns of septin ring localization. Mol Microbiol 41:19–31

Sudbery P (2011) Fluorescent proteins illuminate the structure and function of the hyphal tip apparatus. Fungal Genet Biol 48:849–857

Sutterlin C, Colanzi A (2010) The Golgi and the centrosome: building a functional partnership. J Cell Biol 188:621–628

Taheri-Talesh N, Horio T, Araujo-Bazan L, Dou X, Espeso EA, Penalva MA, Osmani SA, Oakley BR (2008) The tip growth apparatus of *Aspergillus nidulans*. Mol Biol Cell 19:1439–1449

Taheri-Talesh N, Xiong Y, Oakley BR (2012) The functions of myosin II and myosin V homologs in tip growth and septation in *Aspergillus nidulans*. PLoS One 7:e31218

Takeshita N (2016) Coordinated process of polarized growth in filamentous fungi. Biosci Biotechnol Biochem 80:1693–1699

Takeshita N (2018) Oscillatory fungal cell growth. Fungal Genet Biol 110:10–14

Takeshita N, Fischer R (2011) On the role of microtubules, cell end markers, and septal microtubule organizing centres on site selection for polar growth in *Aspergillus nidulans*. Fungal Biol 115:506–517

Takeshita N, Higashitsuji Y, Konzack S, Fischer R (2008) Apical sterol-rich membranes are essential for localizing cell end markers that determine growth directionality in the filamentous fungus *Aspergillus nidulans*. Mol Biol Cell 19:339–351

Takeshita N, Mania D, Herrero de Vega S, Ishitsuka Y, Nienhaus GU, Podolski M, Howard J, Fischer R (2013) The cell end marker TeaA and the microtubule polymerase AlpA contribute to microtubule guidance at the hyphal tip cortex of *Aspergillus nidulans* for polarity maintenance. J Cell Sci 126:5400–5411

Takeshita N, Manck R, Grun N, de Vega SH, Fischer R (2014) Interdependence of the actin and the microtubule cytoskeleton during fungal growth. Curr Opin Microbiol 20:34–41

Takeshita N, Wernet V, Tsuizaki M, Grun N, Hoshi HO, Ohta A, Fischer R, Horiuchi H (2015) Transportation of *Aspergillus nidulans* class III and V chitin synthases to the hyphal tips depends on conventional kinesin. PLoS One 10:e0125937

Takeshita N, Evangelinos M, Zhou L, Serizawa T, Somera-Fajardo RA, Lu L, Takaya N, Nienhaus GU, Fischer R (2017) Pulses of Ca2+ coordinate actin assembly and exocytosis for stepwise cell extension. Proc Natl Acad Sci 114(22):5701–5706

Tan K, Roberts AJ, Chonofsky M, Egan MJ, Reck-Peterson SL (2014) A microscopy-based screen employing multiplex genome sequencing identifies cargo-specific requirements for dynein velocity. Mol Biol Cell 25:669–678

Tatebe H, Nakano K, Maximo R, Shiozaki K (2008) Pom1 DYRK regulates localization of the Rga4 GAP to ensure bipolar activation of Cdc42 in fission yeast. Curr Biol 18:322–330

Torralba S, Raudaskoski M, Pedregosa AM, Laborda F (1998) Effect of cytochalasin A on apical growth, actin cytoskeleton organization and enzyme secretion in *Aspergillus nidulans*. Microbiology 144(Pt 1):45–53

Trinci AP (1974) A study of the kinetics of hyphal extension and branch initiation of fungal mycelia. J Gen Microbiol 81:225–236

Veith D, Scherr N, Efimov VP, Fischer R (2005) Role of the spindle-pole-body protein ApsB and the cortex protein ApsA in microtubule organization and nuclear migration in *Aspergillus nidulans*. J Cell Sci 118:3705–3716

Virag A, Lee MP, Si H, Harris SD (2007) Regulation of hyphal morphogenesis by cdc42 and rac1 homologues in *Aspergillus nidulans*. Mol Microbiol 66:1579–1596

Walther A, Wendland J (2004) Apical localization of actin patches and vacuolar dynamics in *Ashbya gossypii* depend on the WASP homolog Wal1p. J Cell Sci 117:4947–4958

Wang S, Cao J, Liu X, Hu H, Shi J, Zhang S, Keller NP, Lu L (2012) Putative calcium channels CchA and MidA play the important roles in conidiation, hyphal polarity and cell wall components in *Aspergillus nidulans*. PLoS One 7:e46564

Warenda AJ, Konopka JB (2002) Septin function in *Candida albicans* morphogenesis. Mol Biol Cell 13:2732–2746

Wedlich-Soldner R, Bolker M, Kahmann R, Steinberg G (2000) A putative endosomal t-SNARE links exo- and endocytosis in the phytopathogenic fungus Ustilago maydis. EMBO J 19:1974–1986

Wedlich-Soldner R, Altschuler S, Wu L, Li R (2003) Spontaneous cell polarization through actomyosin-based delivery of the Cdc42 GTPase. Science 299:1231–1235

Westfall PJ, Momany M (2002) *Aspergillus nidulans* septin AspB plays pre- and postmitotic roles in septum, branch, and conidiophore development. Mol Biol Cell 13:110–118

Win TZ, Gachet Y, Mulvihill DP, May KM, Hyams JS (2001) Two type V myosins with non-overlapping functions in the fission yeast *Schizosaccharomyces pombe*: Myo52 is concerned with growth polarity and cytokinesis, Myo51 is a cmponent of the cytokinetic actin ring. J Cell Sci 114:69–79

Wollman R, Meyer T (2012) Coordinated oscillations in cortical actin and Ca2+ correlate with cycles of vesicle secretion. Nat Cell Biol 14:1261–1269

Wu CF, Lew DJ (2013) Beyond symmetry-breaking: competition and negative feedback in GTPase regulation. Trends Cell Biol 23:476–483

Xiang X, Fischer R (2004) Nuclear migration and positioning in filamentous fungi. Fungal Genet Biol 41:411–419

Xiong Y, Oakley BR (2009) In vivo analysis of the functions of gamma-tubulin-complex proteins. J Cell Sci 122:4218–4227

Yamashita RA, May GS (1998) Constitutive activation of endocytosis by mutation of myoA, the myosin I gene of *Aspergillus nidulans*. J Biol Chem 273:14644–14648

Zander S, Baumann S, Weidtkamp-Peters S, Feldbrügge M (2016) Endosomal assembly and transport of heteromeric septin complexes promote septin cytoskeleton formation. J Cell Sci 129:2778–2792

Zekert N, Fischer R (2009) The *Aspergillus nidulans* kinesin-3 UncA motor moves vesicles along a subpopulation of microtubules. Mol Biol Cell 20:673–684

Zhang J, Tan K, Wu X, Chen G, Sun J, Reck-Peterson SL, Hammer JA 3rd, Xiang X (2011) *Aspergillus* myosin-V supports polarized growth in the absence of microtubule-based transport. PLoS One 6:e28575

Zhang Y, Gao X, Manck R, Schmid M, Osmani AH, Osmani SA, Takeshita N, Fischer R (2017) Microtubule-organizing centers of *Aspergillus nidulans* are anchored at septa by a disordered protein. Mol Microbiol 106:285–303

Zhou L, Evangelinos M, Wernet V, Eckert AF, Ishitsuka Y, Fischer R, Nienhaus GU, Takeshita N (2018) Superresolution and pulse-chase imaging reveal the role of vesicle transport in polar growth of fungal cells. Sci Adv 4:e1701798

Developmental Decisions in *Aspergillus nidulans*

HEE-SOO PARK[1], MI-KYUNG LEE[2], KAP-HOON HAN[3], MIN-JU KIM[1], JAE-HYUK YU[4,5]

CONTENTS

I. Introduction

Filamentous fungi are ubiquitous eukaryotic microorganisms in nature. Fungi are the main decomposers of organic materials, important environmental nutrient recyclers, and key industrial producers, providing benefits to humankind (Nevalainen and Peterson 2014; Park et al. 2017; Treseder and Lennonb 2015). Conversely, several fungi, such as *Aspergillus flavus* and *Fusarium graminearum*, are key mycotoxin producers that cause a global loss of agricultural commodities (Gugnani 2003; Keller et al. 2005). In addition, a variety of filamentous fungi can cause plant, animal, and human diseases that impact human health and food supplies (Fisher et al. 2012; van Burik and Magee 2001). Owing to the importance of filamentous fungi for humanity, understanding the fungal growth and development would help us to minimize damage and maximize benefits.

Hyphae, long and branching vegetative structures, are the main morphological forms of filamentous fungi in nature (Harris 2011; Riquelme 2013). Fungal cells can undergo reproduction asexually and/or sexually in response to environmental as well as endogenous genetic cues, and these abilities are called **developmental competence** (Axelrod et al. 1973; Noble and Andrianopoulos 2013). In many filamentous fungi, aerial hyphae generally form asexual reproductive structures, the most common reproductive form (Adams et al. 1998). All *Aspergillus* fungi form asexual spores (**conidia**) as the main propagules and infectious particles (Ebbole 2010). Along with the asexual development, some fungi can also reproduce by sexual means and forms sexual structures (Dyer and O'Gorman 2012; Schoustra et al. 2010). These asexual and/or sexual structures (size, shape, color, and arrangement of asexual spores) of filamentous fungi are used for classification (Samson et al. 2014). The formation of asexual and sexual structures is highly sophisticated and regulated by various positive and negative genetic elements that act in several differential stages (Adams et al. 1998; Dyer and O'Gorman 2012). Among filamentous fungi, *Aspergillus*

[1] School of Food Science and Biotechnology, Institute of Agricultural Science and Technology, Kyungpook National University, Daegu, Republic of Korea; e-mail: phsoo97@knu.ac.kr
[2] Biological Resource Center (BRC), Korea Research Institute of Bioscience and Biotechnology (KRIBB), Jeongeup-si, Republic of Korea; e-mail: miklee1010@kribb.re.kr
[3] Department of Pharmaceutical Engineering, Woosuk University, Wanju, Republic of Korea; e-mail: khhan@woosuk.ac.kr
[4] Department of Bacteriology, University of Wisconsin, Madison, WI, USA; e-mail: jyu1@wisc.edu
[5] Department of Systems Biotechnology, Konkuk University, Seoul, Republic of Korea

Biology of the Fungal Cell, 3rd Edition
The Mycota VIII
D. Hoffmeister, M. Gressler (Eds.)

nidulans has been used a model organism to understand developmental biology including fungal growth, conidiation, and sexual differentiation (Casselton and Zolan 2002). This chapter summarizes up-to-date information about regulatory elements and decisions of asexual and sexual development in *A. nidulans.*

II. Developmental Morphology of *Aspergillus nidulans*

Life cycle of *A. nidulans* begins with germination of asexual or sexual spores (Noble and Andrianopoulos 2013). Fungal spores are the most widespread fungal structures in the air (Ebbole 2010) and start to germinate in response to appropriate cues forming the germ tubes (d'Enfert 1997). After germination, the germ tube can further extend apically and form the hyphae, the tube-like structures that are the main mode of vegetative growth (Harris 2011). The hyphae consist of several septated cells and the apical **Spitzenkörper**, the supply center for hyphal tip extension (Harris 2006; Steinberg 2007). During hyphal growth, cytokinesis, septum formation, biosynthesis of cell wall components, and extension of the plasma membrane occur, and these processes are tightly regulated (Harris 2008; Lew 2011). The hyphae must acquire the competence to enter developmental processes. The hyphal cells that have acquired developmental competence cease growth and turn on the developmental programs depending on various environmental stimuli including light, nutrients, oxygen supply, fungal pheromones, and stress conditions (Axelrod et al. 1973; Yager et al. 1982). To obtain the developmental competence from a single spore, approximately 18 and 24 hours of vegetative growth are required for asexual and sexual development, respectively (Axelrod et al. 1973; Noble and Andrianopoulos 2013). Formation of asexual and sexual structures will be completed in about 12 and 27 hours after developmental induction (Fig. 1).

Conidiophores are the asexual developmental structures that bear conidia (Adams et al. 1998; Yu 2010). Development of conidiophores starts with the formation of thick-walled foot cells. The foot cells with developmental competence branch to form aerial stalks under air-exposed conditions. The stalk tip then begins to swell and forms a multinucleate structure called

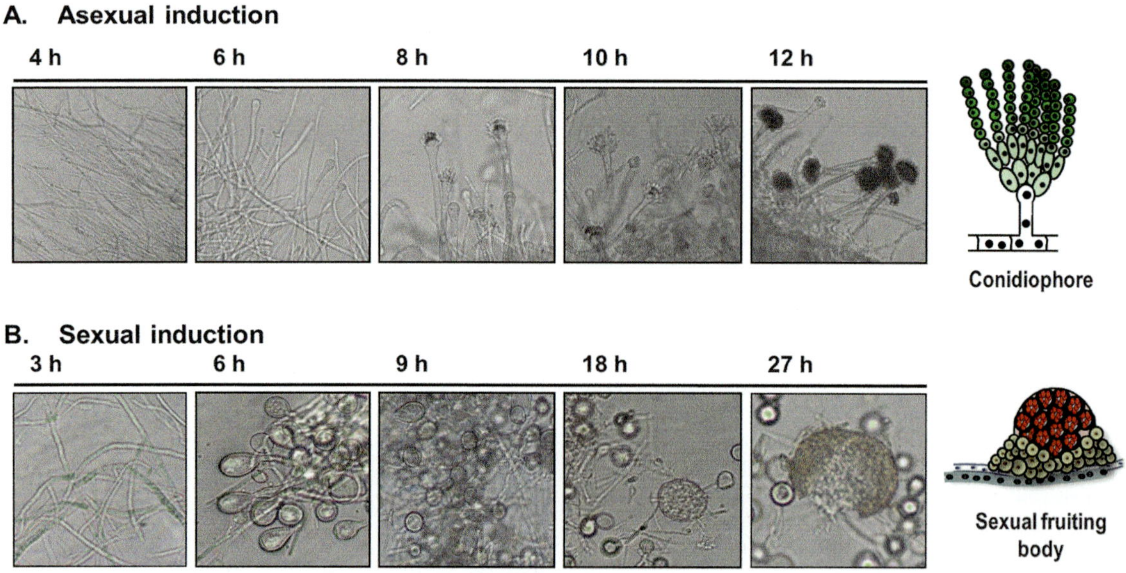

Fig. 1 Developmental processes in *A. nidulans.* (**a**) Conidiophores formation in a wild-type strain after asexual induction. (**b**) Sexual fruiting bodies formation after sexual induction. Dr. Dong-Min Han (Wonkwang University, S. Korea) kindly provided the photomicrographs

vesicle. On the surface of the vesicle, budding-like division occurs, leading to formation of **metulae** and **phialides**, two layers of sterigmata. A secondary layer of sterigmata, termed phialides, generates conidial chains via repeated asymmetric mitotic cell division. After formation of fresh conidia, they undergo maturation processes and finally complete asexual development (Fig. 1a) (Ni et al. 2010; Timberlake 1990; Yu 2010).

Many Aspergilli are able to produce sexual spores with or without mating partners. Some *Aspergillus* species, including *A. nidulans*, are homothallic so they can undergo self-sexual reproduction (Dyer and O'Gorman 2011; Geiser 2009) (Fig. 1b). In *A. nidulans*, sexual reproduction begins with the formation of coiled lumps formed antheridia-like exterior hyphae with core cells (Sohn and Yoon 2002). These coiled lumps are enlarged to form ascogenous hyphae and finally to develop an ascocarp, the fruiting body of ascomycetes, that contains ascospores. In *A. nidulans*, ascocarps are named **cleistothecia** (Krijgsheld et al. 2013). In the early stage of cleistothecia development, the thick-walled globose **Hülle cells** appear, and these cells nurse cleistothecia during sexual development. Young cleistothecia are surrounded by Hülle cells, numerous aerial hyphae, and conidial balls and form a bird nest-like structure; thereby, the species name "*nidulans*" was given (Han 2009; Scherer and Fischer 1998). Core cells of cleistothecia are enlarged and multinucleate, which then form ascogenous cells which then produce the asci (Sohn and Yoon 2002).

III. Environmental Factors Affecting Developmental Fate in *A. nidulans*

The hyphal cells that have acquired developmental competence can undergo asexual or sexual development dependent on the environmental conditions (Riquelme 2013). Various environmental factors, such as nutritional status, air, and culture conditions, control the developmental fate in *A. nidulans* (Bahn et al. 2007; Han et al. 2003; Rai et al. 1967).

Light is one of the most critical factors affecting fungal growth and development (Rodriguez-Romero et al. 2010; Tisch and Schmoll 2010). In the light, *A. nidulans* undergoes asexual development (Mooney and Yager 1990). Light controls expression of certain genes associated with fungal development such as *brlA* and *fluffy* genes (Bayram et al. 2016; Mooney and Yager 1990; Ruger-Herreros et al. 2011; Sarikaya Bayram et al. 2010). *A. nidulans* contains several light sensors that work with the *velvet* regulators and induce mRNA expression of conidiation-specific genes (Bayram et al. 2016; Blumenstein et al. 2005). Three **photoreceptors**, FphA (*f*ungal *ph*ytochrome A), LreA, and LreB (*l*ight *re*sponse A and B), play differential roles in conidiation (Atoui et al. 2010; Blumenstein et al. 2005; Purschwitz et al. 2008, 2009). The deletion of *fphA* encoding a red-light receptor causes reduced *brlA* expression and conidial production, indicating that FphA functions as an activator of conidiation (Atoui et al. 2010; Blumenstein et al. 2005; Ruger-Herreros et al. 2011). Production of conidia in the *lreA* and *lreB* deletion mutants was slightly increased, suggesting that the LreA and LreB complex acts as a repressor of conidiation (Purschwitz et al. 2008; Ruger-Herreros et al. 2011). Light also regulates the localization of the **velvet protein** VeA (*ve*lvet A), a key regulator for development and secondary metabolism in *Aspergillus* spp. (Kim et al. 2002; Stinnett et al. 2007).

Interestingly, VeA interacts with the LreA/LreB/FphA complex and forms the LreA/LreB/FphA/VeA complex, the major light-sensing unit (Bayram et al. 2010; Ruger-Herreros et al. 2011). In general, *A. nidulans* favors sexual development under dark conditions. However, fungal development occurs differentially depending on the light sources. For example, exposure to red or blue light leads to inhibition of sexual development, whereas far-red light can induce sexual development (Bayram et al. 2010; Blumenstein et al. 2005). Light can inhibit sclerotial development in *A. flavus* and *A. parasiticus* (Bennett et al. 1978; Calvo et al. 2004; Duran et al. 2007).

Nutrient sources are also major factors for balancing between asexual and sexual development (Atoui et al. 2010; Han et al. 2003). First, the amount and types of carbon sources affect sexual development (Han et al. 2003). At concentrations of less than 0.5% or higher than 6% glucose, the number of cleistothecia dramatically decreases, suggesting that a certain level of carbon is required for sexual development (Han et al. 2003). Several carbon sources such as lactose and glycerol favor sexual development, whereas acetate can block formation of cleistothecia. Second, the type of nitrogen sources is important for deciding developmental process, and organic nitrogen sources can induce sexual development in *A. nidulans* (Han et al. 2003). The ratio between carbon and nitrogen is most important for asexual or sexual reproduction (Han et al. 2003). Nutrient starvations can induce fungal development in the submerged culture (Martinelli 1976; Saxena and Sinha 1973). Carbon and nitrogen starvation induces *brlA* expression and causes asexual developmental induction in differential pattern (Skromne et al. 1995). Glucose starvation causes formation spores on abnormal conidiophores that bypass the vesicle and metulae stages, whereas nitrogen starvation induces the production of more elaborate conidiophores (Skromne et al. 1995).

A study proposes that FlbD (*fluffy low* brlA *locus D*), a cMyb-type transcription factor (TF) necessary for the proper expression of *brlA*, is associated in response to nitrogen starvation (Arratia-Quijada et al. 2012). High concentrations of salts, such as sodium chloride or potassium chloride, activate conidiation (Han et al. 2003; Lee and Adams 1994).

Oxygen also affects fungal development (Grahl et al. 2012). Fungal hyphae grow in submerged culture condition or in restricted exposure to air. After acquired developmental competence, a high oxygen concentration (air exposure) can induce the production of conidiophores (Adams et al. 1998; Axelrod et al. 1973; Morton 1961). Conversely, low concentrations of oxygen can initiate sexual development (Zonneveld 1988). However, the detailed mechanisms of fungal development regulated by air remain to be understood. Osmolarity is also responsible for the preferential development of conidia (Lee and Adams 1994). Addition of 1 M KCl or 1 M NaCl can induce production of asexual spores but decrease sexual development. However, higher concentration of salts can inhibit fungal growth, blocking both asexual and sexual development (Han et al. 2003; Song et al. 2001).

IV. Developmental Decision for Conidiation

Conidiation in *Aspergillus* occurs as an integral part of the life cycle primarily controlled by the intrinsic genetic program. The formation of conidiophore is tightly regulated by multiple genetic elements, and these are extensively studied in *A. nidulans* (Fig. 2) (Adams et al. 1998; Park and Yu 2012). Three TFs BrlA (*br*ist*l*e A), AbaA (*aba*cus A), and WetA (*wet*-white A) are central regulators for conidiation that control expression of genes associated with the assembly of the conidiophore (Adams et al. 1998; Yu 2010). To activate central regulators of asexual development, upstream regulators should induce *brlA* expression, and several **repressors** should be removed from the promoter regions of *brlA* (Lee et al. 2016) (Fig. 2c). A recent study has revealed that there are at least three negative regulators of conidiation and that a key event for the acquisition of the asexual developmental competence is to remove the repressive effects imposed by NsdD (*n*ever in *s*exual *d*evelopment locus *D*) and VosA (*v*iability *o*f *s*pores A) (Lee et al. 2016). Importantly, for the first time, this study demonstrated that NsdD physically binds to three different regions in the *brlA*β promoter, further supporting the idea that NsdD directly (rather than indirectly) represses the onset of *brlA*β expression and conidiation. After completion of conidiophore formation, feedback regulators turn off the activities of the central regulators (Ni et al. 2010). The VosA-VelB (*vel*vet-*l*ike B) complex acts as a key feedback regulator that represses *brlA* expression on conidia (Ni and Yu 2007; Park et al. 2012).

A. Upstream Regulators of Conidiation

In response to environmental conditions, hyphal cells that have acquired developmental competence stop vegetative growth and start conidiation (Adams et al. 1998). Various studies have identified several upstream developmental regulators that induce *brlA* expression. Mutations in any of the genes *fluG* (*fluffy* locus A) and *flbA-E* (*fluffy low* brlA loci A~E) lead to

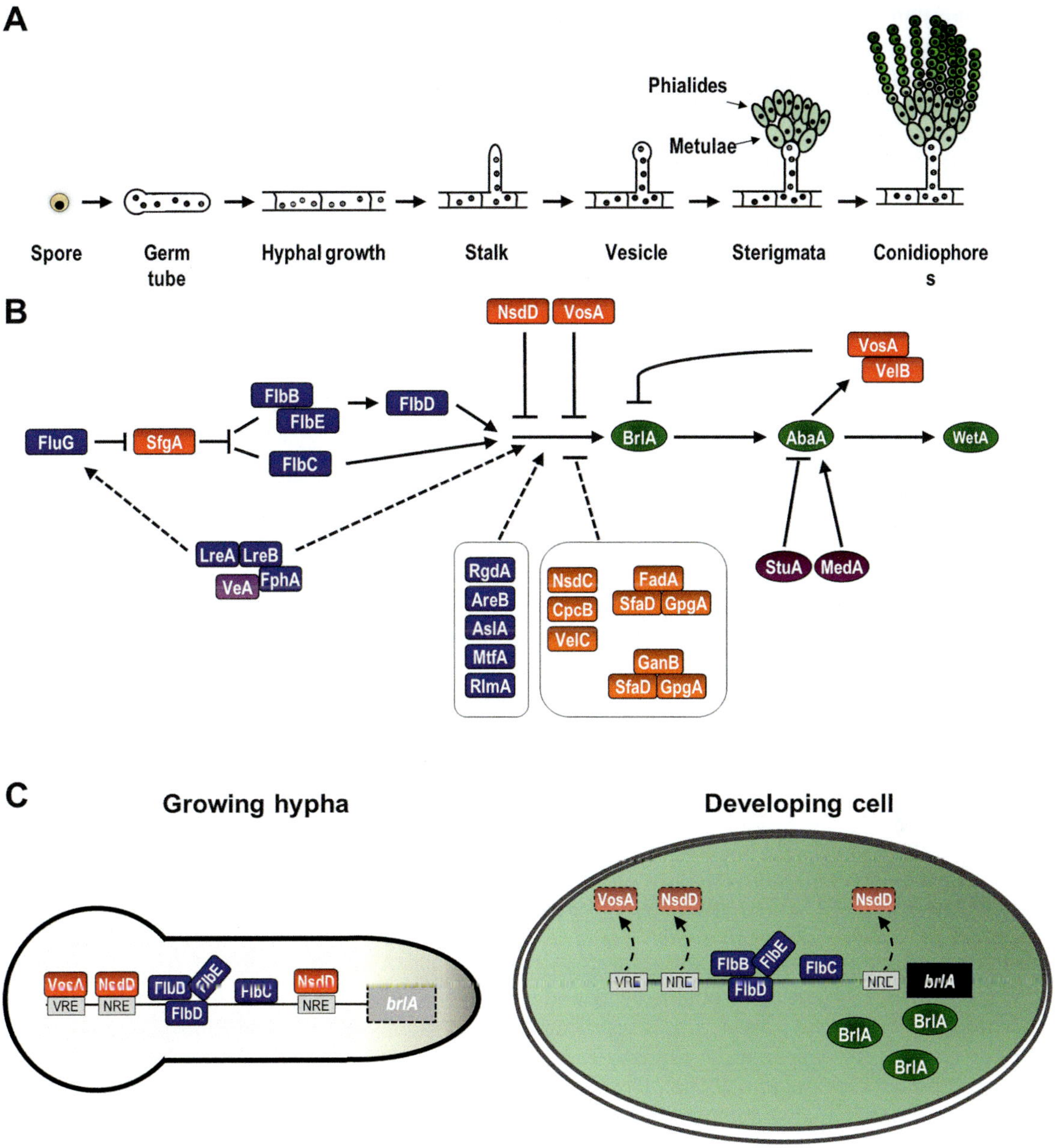

Fig. 2 Asexual development in *A. nidulans*. (**a**) A schematic presentation of morphological changes during conidiophore formation in *A. nidulans*. (**b**) A genetic model for developmental regulation. Several upstream activators are required for initiation of central regulator pathway (BrlA → AbaA → WetA) of asexual development. In contrast, several TFs, including SfgA, VosA, and NsdD, or the G protein signaling pathways, repress initiation of asexual development. (**c**) A model depicting the roles of positive and negative regulators in governing the acquisition of the developmental competence. In hyphae, VosA and/or NsdD bind(s) to the upstream regulator region of *brlA*, which then repress of mRNA expression of *brlA*. In developing cells, NsdD and VosA can be displaced from the *brlA* promoter, and the FlbB-FlbD and FlbC activate *brlA* expression and conidiation

"fluffy," cotton-like phenotypes (Adams et al. 1998; Etxebeste et al. 2010).

FluG is one of the upstream developmental activators which is required for inhibition of vegetative growth and initiation of conidiation

(Lee and Adams 1994). Overexpression of *fluG* causes conidiophore formation and *brlA* activation in liquid submerged cultures, whereas the deletion of *fluG* leads to the fluffy phenotypes (D'Souza et al. 2001; Lee and Adams 1996; Wieser et al. 1994). FluG is essential for synthesis of a diorcinol-dehydroaustinol adduct, an extracellular sporulation inducing factor (known as the **FluG factor**), which signals the activation of conidiophore development (Lee and Adams 1994; Rodriguez-Urra et al. 2012). This adduct can rescue asexual developmental defects caused by absence of *fluG* (Rodriguez-Urra et al. 2012). FluG-mediated signaling regulates proliferation and development via two independent pathways; the cessation of vegetative growth via FlbA activation and the initiation of conidiation via activation of developmental genes (*flbB~E*) (Yu 2010). The FluG-mediated developmental regulation is divided in two independent pathways, FlbE/FlbB/FlbD and FlbC, for initiation of conidiation and activation of *brlA* (Etxebeste et al. 2010; Park and Yu 2012).

Four upstream transcriptional activators including FlbB, FlbC, FlbD, and FlbE regulated by FluG are putative TFs which are needed for cessation for hyphal cell growth and regulation of development (Park and Yu 2012). FlbC contains two C_2H_2 zinc finger DNA-binding domains which are required for interaction of the promoter region of *brlA* and activation of *brlA* (Kwon et al. 2010a). Deletion of *flbC* leads to reduction in conidiation, whereas overexpression of *flbC* causes inhibition of hyphal growth and induction of *brlA*, *abaA*, and *vosA*, suggesting that FlbC is vital for coordinating fungal growth and development (Kwon et al. 2010a). FlbB has a basic leucine zipper (b-zip) domain and localizes at the hyphal tip (Etxebeste et al. 2009). FlbB interacts with FlbE and forms the FlbB-FlbE complex which activates *flbD* expression (Etxebeste et al. 2009; Garzia et al. 2009; Kwon et al. 2010a). Then, FlbD, a cMyb-type TF, also forms a complex with FlbB in the nucleus, and this complex directly binds to the promoter region of *brlA* and activates its expression (Garzia et al. 2010).

In addition to the roles of FlbB~FlbEs for growth and asexual development, several studies demonstrated that FlbB~FlbE are also required for proper sexual development (Arratia-Quijada et al. 2012; Kwon et al. 2010a, b; Oiartzabal-Arano et al. 2015).

Proper and precise control of *brlA* in vegetative cells is extremely crucial for the survival and fitness of *Aspergillus* fungi (Lee et al. 2016). During the early phase of vegetative growth, the Flb proteins can occupy in the promoter region of *brlA*. However, the Flb proteins cannot induce *brlA* transcription, as several repressors directly bind to the *brlA* promoter and interfere with the function of the Flb proteins (Lee et al. 2016). Gain-of-function genetic screens proposed that VosA acts as a key repressor of conidiation (Ni and Yu 2007). In hyphal cells, VosA, a fungal-specific velvet family TF, forms the VosA-VosA homodimer or the VosA-VelB heterodimer and represses *brlA* expression in liquid submerged culture (Park et al. 2012; Sarikaya Bayram et al. 2010). NsdD is another regulator that represses *brlA* expression during vegetative growth (Lee et al. 2014). In the developing cell, the NsdD and VosA proteins may be subject to degradation and removed from the *brlA* promoter, and then the Flb proteins induce *brlA* expression (Lee et al. 2016) (Fig. 2c).

Several TFs have been shown to influence growth in *A. nidulans*. SfgA is a putative TF with a $Zn(II)_2Cys_6$ binuclear DNA-binding domain (Seo et al. 2003, 2006). SfgA plays a role downstream of FluG but upstream of FlbA, FlbB, FlbC, FlbD, and BrlA (Seo et al. 2006). OsaA (*o*rchestrator of *s*ex and *a*sex A) is a functional equivalent of Wor1 in *Candida albicans*. The presence of multiple *osaA* copies in its genome represses conidiation (Alkahyyat et al. 2015; Ni and Yu 2007). RgdA (a putative APSES TF), MtfA (*m*aster *t*ranscription *f*actor A; a C_2H_2 zinc finger TF), RlmA (a major MpkA-dependent TF), AreB (a putative GATA zinc-finger TF), and AslA (*a*sexual differentiation with *l*ow-level conidiation A; a C_2H_2-type zinc finger TF) are also involved in normal growth and development in *A. nidulans* (Kim et al. 2017; Kovacs et al. 2013; Lee et al. 2013; Ramamoorthy et al. 2013; Wong et al. 2009). De Souza and colleagues identified various kinases, such as CkiB, Gsk3, PkaA, NikA, and PlkA, which are essential for proper fungal growth (De Souza et al. 2013).

Heterotrimeric G proteins (G proteins) compose of α, β, and γ subunits that are involved in most biological processes in filamentous fungi (Yu 2006). In *A. nidulans*, two heterotrimeric G protein signaling pathways, FadA (*fluffy autolytic dominant* A)-mediated and GanB (*G* protein *alpha* subunit in *A. nidulans* B)-mediated signaling pathways, were studied, and these two pathways govern fungal growth, development, and secondary metabolism (Chang et al. 2004; Wieser et al. 1994; Yu 2006; Yu et al. 1996). In response to environmental stresses, two G alpha proteins, FadA and GanB, dissociate from the cognate GPCR (G protein-coupled receptor) and the Gβγ hetero-complex SfaD(Gβ)::GpgA(Gγ), and the dissociated Gα subunit (FadA and/or GanB) and/or the Gβγ hetero-complex cooperatively regulates vegetative growth and represses both asexual and sexual development via the cyclic AMP (cAMP)-dependent protein kinase PkaA (Lafon et al. 2005; Rosen et al. 1999; Seo et al. 2005; Shimizu and Keller 2001). FadA- and GanB-mediated signaling pathways are negatively controlled by the *r*egulators of *G* protein *s*ignaling (RGSs) FlbA and RgsA (*r*egulators of *G* protein *s*ignaling A), respectively (Han et al. 2004b; Hicks et al. 1997; Wieser et al. 1997). In addition, GanB-mediated signaling is in part activated by the putative GDP/GTP exchange factor RicA (an orthologue of *Caenorhabditis elegans* RIC-8) (Kwon et al. 2012). Another G protein component CpcB (*c*ross-*p*athway *c*ontrol B; Gβ-like protein B) is required for proper fungal growth and development in *A. nidulans* (Kong et al. 2013).

In fungi, MAPKs (mitogen-activated protein kinases) are involved in hyphal growth, development, and virulence (Xu 2000). Among four MAPK genes, including *mpkA*, *mpkB*, *mpkC*, and *hogA* in *A. nidulans*, *mpkB* encodes a homolog to Fus3p of the baker's yeast and is required for proper fungal growth, development, and secondary metabolism (Atoui et al. 2008; Bayram et al. 2012; Jun et al. 2011; Kang et al. 2013; Paoletti et al. 2007). Deletion of *mpkB* results in increased expression of *brlA* and decreased VeA phosphorylation and VeA-VelB formation, which function as an activator of sexual development, suggesting that MpkB plays an important role in both asexual and sexual development (Bayram et al. 2012; Kang et al. 2013).

B. Initiation of Conidiation

Under appropriate conditions, some of the vegetative cells cease hyphal growth and initiate conidiation (Adams et al. 1998). The key step for developmental transition from apical growth to conidiation is **activation of *brlA*** (Adams et al. 1988, 1990). *brlA* null mutants show phenotypes including indeterminate structures that resemble conidiophore stalks (thus termed "bristle") and fail to form any asexual structures including vesicles, metulae, phialides, and conidia (Adams et al. 1988). In contrast, overexpression of *brlA* leads to termination of hyphal growth and the formation of viable spores from hyphal apices (Adams et al. 1988). External signals, such as nutrient limitations or several stresses, cannot bypass the BrlA requirement for asexual development, suggesting that *brlA* activation is an essential control step for commencing conidiation.

A recent study has revealed that the abovementioned upstream developmental activators are needed for maximum conidiation, but not for the commencement of development. This is based on the fact that the deletion of *nsdD* could bypass the need for *fluG*, *flbB*, *flbE*, *flbD*, and *flbC*, but not *brlA*, in conidiation (Lee et al. 2016).

Once the negative regulators NsdD and VosA are removed and upstream activators maximize expression of *brlA*, the C_2H_2 zinc finger TF BrlA activates expression of several genes involved in conidiation (Adams et al. 1990). Deletion of *brlA* blocks expression of *abaA* and *wetA*, whereas forced expression of *brlA* leads to activation of developmental regulatory genes (Mirabito et al. 1989). These developmental genes, including *abaA*, *wetA*, *rodA*, and *yA*, contain the BrlA response elements (BREs; 5′-(C/A)(G/A)AGGG(G/A)-3′) in their promoter regions (Chang and Timberlake 1993; Prade and Timberlake 1993). The *brlA* locus consists of two overlapping transcriptional units, designated *brlA*α and *brlA*β. The regulatory mechanisms of *brlA*α and *brlA*β are different. *brlA*α is controlled via a transcriptional mechanism, while *brlA*β is regulated at both the transcriptional and translation levels (Han and Adams 2001; Han et al. 1993). The *brlA*β mRNA is produced in vegetative cells before developmental induction, but it does not accumulate to substantial levels, likely because translation of the *brlA*β μORF represses BrlAβ translation to block development. Following BrlAβ translation, *brlA*α transcription is activated primarily through the *brlA*-dependent positive feedback loop (Adams et al. 1998). The ultimate result of *brlA* activation is activation of other development-specific genes including *abaA* and *wetA*.

C. Progression and Termination of Conidiation

After activation of *brlA*, BrlA directly induces expression of *abaA* required for formation of phialides during the middle phase of asexual development (Boylan et al. 1987; Sewall et al. 1990a). The *abaA* null mutant forms non-sporulating conidiophores, similar to an abacus-like structure, and does not form phialides, suggesting that *abaA* is required for proper formation of phialides (Clutterbuck 1969; Sewall et al. 1990a). Overexpression of *abaA* leads to cessation of vegetative growth and accentuates cellular vacuolization without spore formation in liquid submerged culture (Mirabito et al. 1989).

AbaA is a TEF1 (transcriptional enhancer factor-1) family member which contains an ATTS (AbaA, TEC1p, TEF-1 sequence)/TEA DNA-binding motif (Andrianopoulos and Timberlake 1991, 1994). AbaA binds to the *cis* consensus sequence 5′-CATTCY-3′ (AbaA response element (ARE), where Y is a T or C) and regulates their expression during phialide differentiation (Andrianopoulos and Timberlake 1994). Previous studies demonstrated that AbaA positively regulates expression of several genes, including the chitin synthase gene *chsC*, a component of the axial bud site marker *axl2*; developmental genes including *yA*, *rodA*, *wA*, *brlA*, *wetA*, *vosA*, and *velB*; and *abaA* itself, which contain AREs in their promoter regions (Aguirre et al. 1990; Aramayo and Timberlake 1993; Ichinomiya et al. 2005; Park et al. 2003, 2012; Si et al. 2012). In addition, AbaA is required for repression of *brlA* during mid-phage of conidiation, without requiring AbaA binding to the *brlA* promoter region, suggesting that AbaA may indirectly repress *brlA* expression (Aguirre 1993; Han and Adams 2001).

During late phase of conidiation, WetA, VosA, and VelB play crucial roles in formation, maturation, integrity, and dormancy of conidia (Marshall and Timberlake 1991; Ni and Yu 2007; Park et al. 2012; Sewall et al. 1990b). WetA is a key regulator for the conidium wall modification which is essential for the stability of mature and dormant conidia (Marshall and Timberlake 1991; Sewall et al. 1990b).

The *wetA* mutant produces colorless and autolytic conidia, described as "wet-white" (Clutterbuck 1969). In addition, the *wetA* mutant conidia lack of both the condensation of the C2 wall layer and the formation of C3 and C4 layers (Sewall et al. 1990b). WetA also acts as a regulator of conidium-specific genes including *wA* (Marshall and Timberlake 1991). With BrlA and AbaA, WetA has been proposed to define a central regulatory pathway that functions in concert with other genes to regulate conidiation-specific gene expression and determine the order of gene activation and repression (Adams et al. 1998; Mirabito et al. 1989).

In conidia, two velvet regulators VosA and VelB interact with each other and form the VosA-VelB complex that plays a crucial role in conidial maturation, conidial trehalose biogenesis (Ni and Yu 2007; Park et al. 2012; Sarikaya Bayram et al. 2010). The deletion of *vosA* or *velB* results in a loss of conidial viability, the lack of trehalose in conidia, and a reduction of conidial tolerance to environmental stresses (Ni and Yu 2007; Park et al. 2012; Sarikaya Bayram et al. 2010). The velvet regulators are fungal-specific TFs which have the DNA-binding *velvet* motif (Ahmed et al. 2013). The VosA-VelB complex positively regulates the expression of conidia-specific genes and represses certain development-associated genes (Ahmed et al. 2013; Park et al. 2015). Overall, the VosA-VelB complex controls the commencement, progression, and completion of sporogenesis.

Two developmental modifiers, StuA (*stu*nted) and MedA (*med*usa), work with central regulatory genes and are necessary for the precise organization of conidiophores (Adams et al. 1998). StuA is a TF containing the APSES motif and is required for proper activation of *brlA* and repression of *abaA* (Dutton et al. 1997). MedA is also required for proper expression of *brlA* and *abaA* and proper formation of conidiophores (Busby et al. 1996).

V. Developmental Decisions for Sexual Development

Due to the complexity of the sexual reproduction, only a few studies have been conducted compared to conidiation (Dyer and O'Gorman 2012). Like conidiation, *A. nidulans* has several advantages in studying sexual reproduction, including the homothallic sexual cycle, early availability of the whole genome data, and various tools for genetic manipulation; thus it has been used to identify and characterize genes associated with sexual fruiting (Archer and Dyer 2004; Galagan et al. 2005; Todd et al. 2007). Less than 100 genes required for proper sexual development in *Aspergillus* spp. have been identified (Fig. 3), and their roles have been described in the other excellent reviews (Dyer and O'Gorman 2012; Dyer et al. 2003). In this section, functions of select genes are summarized.

Most **homothallic fungi** contain mating-type genes which are crucial for mating processes. In *A. nidulans*, two genes MAT-1 (*matB*) and MAT-2 (*matA*) were the first to be described. They were distinct from other fungi in that they are not localized on the same chromosome (Paoletti et al. 2007). Deletion of *MAT1* or *MAT2* results in a significantly decreased number of abnormal cleistothecia. Overexpression of the mating-type genes causes cleistothecia production on submerged liquid culture which represents unfavorable conditions for sexual development, suggesting that there are key genes for sexual development (Paoletti et al. 2007).

As mentioned above, the velvet family proteins are multifunctional coordinators of fungal growth, conidiation, sexual development, and secondary metabolism in filamentous fungi (Bayram and Braus 2012). The roles of the

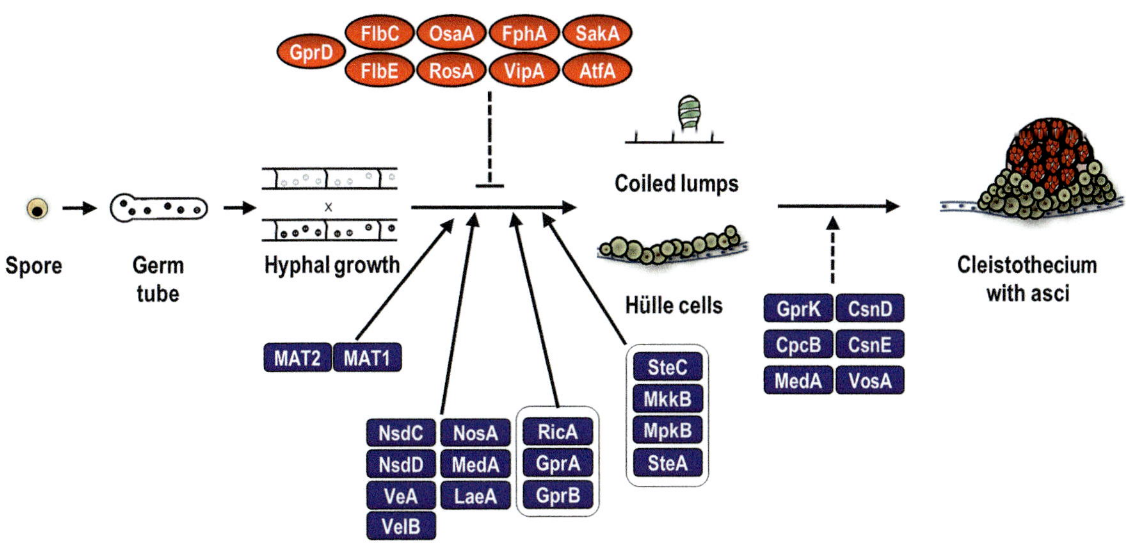

Fig. 3 Sexual development of *A. nidulans*. A schematic presentation of development of sexual fruiting bodies and associated developmental genes in *A. nidulans*. Proteins in the red circle can act as repressors during sexual development. Whereas, proteins in the blue rectangle can induce formation of sexual fruiting bodies. (See main text for details)

velvet proteins in sexual development were described by Käfer (Kafer 1965). The *veA1* mutant produced decreased numbers of sexual fruiting bodies with increased conidiation, suggesting that VeA acts as a balancer between asexual and sexual development (Kafer 1965; Mooney and Yager 1990). Follow-up studies extensively characterized the roles of the velvet proteins. Importantly, Kim et al. identified the *veA* gene and showed that VeA was crucial for sexual development and sterigmatocystin production (Kim et al. 2002). Later studies have revealed the molecular mechanisms of VeA action, which showed that the binding partners and localization of VeA were important for the roles of VeA in sexual development, especially the light-dependent condition (Bayram et al. 2008; Stinnett et al. 2007). In the light, VeA is mainly localized in the cytoplasm; therefore it cannot induce sexual development. The nuclear localization of VeA is regulated by the light complex components FphA, LreA, and LreB (Purschwitz et al. 2008, 2009). Under dark conditions, however, VeA interacts with VelB in the cytoplasm and translocates in the nucleus, leading to the formation of the VeA-VelB or VelB-VelB-LaeA complexes that control cleistothecia production and sterigmatocystin biosynthesis (Bayram et al. 2008). Deletion of either *veA* or *velB* results in the absence of sexual fruiting bodies under sexually favorable conditions (Kim et al. 2002; Park et al. 2012). The *laeA* deletion mutant also produced abnormal cleistothecia (Sarikaya Bayram et al. 2010). Another VeA interacting protein, VipA (*v*eA-*i*nteracting *p*rotein A), is also involved in the light-dependent developmental process. The phenotype of the *vipA* null mutant is similar to that of the *fphA* null mutant (Rohrig et al. 2017). Recently, Rauscher et al. demonstrated that phosphorylation of VeA affects their roles in sexual development (Rauscher et al. 2016). With VeA, VelB is also required for the initiation of sexual development.

VelB exists in both the VeA-VelB and VelB-VosA complexes that play different roles in hyphae. Previously, we proposed the ratio of VeA-VelB and VelB-VosA is crucial for initiation of sexual reproduction, and this ratio can be regulated by VelC (*v*e*l*vet-*l*ike C). After acquisition of the sexual developmental competence, VelC is produced and forms the VosA-VelC complex, leading to decreased formation of the VelB-VosA hetero-complex whereas increased formation of VelB-VeA (Park et al. 2014).

Classical genetic approaches are a useful way to identify genes that play crucial roles in developmental stages (Han 2009). Han and colleagues screened massive mutants showing defective sexual reproduction and classified them into three groups: **NSD mutants** (never in sexual development), **BSD mutants** (block in sexual development), and **ASD mutants** (abnormal in sexual development) (Han et al. 1990). NSD mutants exhibited common phenotypes including the absence of sexual fruiting bodies, apical growth, and earlier development of conidiospores (Han et al. 1994, 1998). Among them, two genes, *nsdC* and *nsdD*, were further characterized (Han et al. 2001; Kim et al. 2009). NsdC contains a C_2H_2-C_2H_2-C_2HC zinc finger DNA-binding domain and acts as a key positive regulator of sexual development. Overexpression of *nsdC* leads to increased formation of sexual fruiting bodies and overcome environmental factors which inhibit cleistothecial development (Kim et al. 2009). NsdC is also required for repression of asexual development and *brlA* expression, suggesting that NsdC can regulate balance between asexual and sexual development. NsdD is a GATA-type TF that functions as activator of sexual reproduction (Han et al. 2001). Similar to the *nsdC* mutants, strains overexpressing *nsdD* can produce more sexual fruiting bodies compared to the wild-type strain (Han et al. 2001).

Signal transduction pathways, including G protein signaling pathways and **mitogen-activated protein kinase** (MAPK) cascades, play multifunctional roles in biological processes in most organisms (Lengeler et al. 2000). In *A. nidulans*, several G protein-coupled receptors have been shown to be required for self-fertilization, sexual reproduction, and/or secondary metabolism (Han et al. 2004a; Seo et al. 2005). Three G protein-coupled receptors (Gpr), GprA, GprB, and GprD, are required for proper sexual development (Han et al. 2004a; Seo et al. 2004). Especially, GprD signaling pathway may act as an upstream negative regulator for GprA- and GprB-mediated sexual reproduction and the formation of sexual fruiting bodies (Han et al. 2004a; Seo et al. 2004). Another G protein-coupled receptor, GprK, might be required for the maturation of cleistothecia. The *gprK* deletion mutant can produce Hülle cells but is blocked in formation cleistothecia (Dyer and O'Gorman 2012). The Gβ-like protein CpcB is involved in middle of end phase of sexual development (Hoffmann et al. 2000; Kong et al. 2013; Palmer et al. 2006). Deletion of *cpcB* causes decreased formation of cleistothecia and the absence of ascospores in cleistothecia (Kong et al. 2013). The guanine nucleotide exchange factor RicA regulates the G protein signaling pathway required for fungal growth and sexual development. The absence of *ricA* resulted in the lack of Hülle cells or cleistothecia formation (Kwon et al. 2012). Overall, these results demonstrate that several G protein signaling pathways play a crucial role in the sexual development in *A. nidulans*.

As mentioned above, the functions of MAPK pathways were well characterized in *Saccharomyces cerevisiae* and other filamentous fungi, and several kinases in the MAPK signaling cascade are involved in pheromone response, pathogenesis, and stress responses (Xu 2000). SteC (equivalent of yeast Ste11p, MAPKKK, or MAPKK kinase), MkkB (homolog of yeast Ste11p, MAPKK, or MAPK kinase), MpkB (homolog of yeast Fus3p, MAPK, or MAP kinase), and SteA (homolog of yeast Ste12p) are components of MAPK cascade in *A. nidulans*, and these proteins work together to regulate sexual development (Bayram et al. 2012; Paoletti et al. 2007; Vallim et al. 2000; Wei et al. 2003). The absence of any of these genes results in failure to form ascogenous hyphae and cleistothecia. The **HOG pathway** (*h*igh-*o*smolarity *g*lycerol) is mainly involved in stress response in many yeast and fungi. However, this pathway is also involved in controlling sexual development in *A. nidulans* (Kawasaki et al. 2002). Kawasaki and colleagues found that deletion of *sakA* (*hogA*) causes increased production of cleistothecia, suggesting that SakA (*s*tress *a*ctivated *k*inase A) acts as a repressor of sexual development (Kawasaki et al. 2002). The SakA interacting protein AtfA (homolog of *Schizosaccharomyces pombe* Atf1) is also associated with sexual development as the *atfA* deletion mutant produces increased number of cleistothecia (Lara-Rojas et al. 2011).

The balance between asexual and sexual development is regulated by various factors. Three oxylipin biosynthetic genes *ppoA*, *ppoB*, and *ppoC* (psi factor producing oxygenase) are required for proper asexual and sexual development (Brodhun and Feussner 2011; Tsitsigiannis and Keller 2007; Tsitsigiannis et al. 2004, 2005). Deletion of *ppoA* or *ppoB* causes increased conidial production suggesting that PpoA and PpoB negatively affect asexual development (Tsitsigiannis et al. 2004, 2005). However, deletion of *ppoC* leads to decreased asexual sporulation, suggesting that PpoC positively influences conidiation and antagonizes PpoA and PpoB (Tsitsigiannis et al. 2004, 2005). As mentioned above, OsaA functions as a key orchestrator of sexual and asexual development (Alkahyyat et al. 2015). The deletion of *osaA* causes enhanced sexual fruiting with reduced conidiation, suggesting that OsaA acts as a repressor of sexual development and indirectly affects asexual development in a positive way. The deletion of *osaA* could suppress the *veA1* mutant allele leading to the sexual developmental phenotype similar to that of *veA*+ wild type. This indicates that OsaA acts downstream of VeA as a key negative regulator of sexual development (Alkahyyat et al. 2015). Thus, one key event to achieve the sexual developmental competence is to remove the repressive effects imposed by OsaA.

Some asexual regulators are involved in both asexual and sexual development. StuA and MedA are asexual developmental modifiers that are necessary for formation of cleistothecia (Clutterbuck 1969; Martinelli 1976). While the *stuA* deletion mutant cannot produce Hülle cells, the *med1* mutant is able to produce Hülle cells, suggesting that these two modulators play different roles in sexual development (Busby et al. 1996; Wu and Miller 1997). Upstream activators of asexual development FlbC and FlbE repress sexual development. Deletion of either *flbC* or *flbE* resulted in increased formation of cleistothecia (Kwon et al. 2010a, b).

CSN, the **constitutive photomorphogenesis complex 9** (COP9) **signalosome,** is a multi-subunit complex that is involved in multiple fungal developmental processes (Braus et al. 2010). In *A. nidulans*, eight subunits were identified, and these subunits play diverse roles in sexual development. For example, the *csnD* deletion mutant cannot enter the primordial stage in sexual development, but this mutant can produce primordia under light conditions (Busch et al. 2003). CsnE is required for expression of cell wall-degrading enzymes and maturation of sexual fruiting bodies (Nahlik et al. 2010). Unlike CsnD and CsnE, three subunits including CsnA, CsnB, and CsnG act as activators of sexual development (Busch et al. 2003, 2007; Nahlik et al. 2010). Overall, CSN is a major contributor to regulate sexual developmental processes (Braus et al. 2010).

Two orthologues of *Sordaria macrospora* Pro1, RosA (*r*epressor *o*f *s*exual development) and NosA (*n*umber *o*f *s*exual spores), were found in the *Aspergillus* genome (Vienken and Fischer 2006; Vienken et al. 2005). Genetic analysis proposed that NosA might be an NsdD downstream activator for sexual development and be associated with a completion of sexual reproduction (Vienken and Fischer 2006). Unlike NosA, RosA regulates expression of genes involved in sexual primordia and, hence, functions in an early stage of sexual development (Vienken et al. 2005).

VI. Conclusions

Fungal development is a very complex process which is influenced by various internal and/or external factors. To enter development stages, fungi must acquire the developmental competence. Increasing evidence from numerous recent investigations suggest that the key event for the acquisition of the developmental competence is to remove the repressive effects imposed by multiple negative regulators of asexual or sexual development. It appears that, even in the presence of developmental activators, the initiation of developmental processes would not occur, as long as developmental repressors or growth inducers are prevailing. During developmental stages, several TFs or signal cascades control developmental processes. In this chapter, we have summarized key regulators for making the developmental decisions and their roles in sexual and asexual development in *A. nidulans*. Further studies aimed at revealing the detailed molecular mechanisms of sexual or asexual reproduction in diverse fungal species will illuminate the common and distinct regulators and signaling cascades governing growth and development in fungi.

Acknowledgments The work by HSP and MJK was supported by the National Research Foundation of Korea (NRF) grant to HSP funded by the Korean government (MSIP: No. 2016010945). The work by KHH was supported by the Intelligent Synthetic Biology Center of Global Frontier Projects (2015M3A6A8065838) and by Basic Science Research Program through NRF (NRF-2017R1D1A3B06035312) funded by Korean government. The work by MKL and JHY was supported by the Intelligent Synthetic Biology Center of Global Frontier Project (2011-0031955) funded by the Ministry of Education, Science and Technology grants.

References

Adams TH, Boylan MT, Timberlake WE (1988) *brlA* is necessary and sufficient to direct conidiophore development in *Aspergillus nidulans*. Cell 54:353–362

Adams TH, Deising H, Timberlake WE (1990) *brlA* requires both zinc fingers to induce development. Mol Cell Biol 10:1815–1817

Adams TH, Wieser JK, Yu J-H (1998) Asexual sporulation in *Aspergillus nidulans*. Microbiol Mol Biol Rev 62:35–54

Aguirre J (1993) Spatial and temporal controls of the *Aspergillus brlA* developmental regulatory gene. Mol Microbiol 8:211–218

Aguirre J, Adams TH, Timberlake WE (1990) Spatial control of developmental regulatory genes in *Aspergillus nidulans*. Exp Mycol 14:290–293

Ahmed YL, Gerke J, Park H-S, Bayram O, Neumann P, Ni M, Dickmanns A, Kim SC, Yu J-H, Braus GH et al (2013) The *velvet* family of fungal regulators contains a DNA-binding domain structurally similar to NF-κB. PLoS Biol 11:e1001750

Alkahyyat F, Ni M, Kim SC, Yu JH (2015) The WOPR domain protein OsaA orchestrates development in *Aspergillus nidulans*. PLoS One 10:e0137554

Andrianopoulos A, Timberlake WE (1991) ATTS, a new and conserved DNA binding domain. Plant Cell 3:747–748

Andrianopoulos A, Timberlake WE (1994) The *Aspergillus nidulans abaA* gene encodes a transcriptional activator that acts as a genetic switch to control development. Mol Cell Biol 14:2503–2515

Aramayo R, Timberlake WE (1993) The *Aspergillus nidulans yA* gene is regulated by *abaA*. EMBO J 12:2039–2048

Archer DB, Dyer PS (2004) From genomics to post-genomics in *Aspergillus*. Curr Opin Microbiol 7:499–504

Arratia-Quijada J, Sanchez O, Scazzocchio C, Aguirre J (2012) FlbD, a Myb transcription factor of *Aspergillus nidulans*, is uniquely involved in both asexual and sexual differentiation. Eukaryot Cell 11:1132–1142

Atoui A, Bao D, Kaur N, Grayburn WS, Calvo AM (2008) *Aspergillus nidulans* natural product biosynthesis is regulated by *mpkB*, a putative pheromone response mitogen-activated protein kinase. Appl Environ Microbiol 74:3596–3600

Atoui A, Kastner C, Larey CM, Thokala R, Etxebeste O, Espeso EA, Fischer R, Calvo AM (2010) Cross-talk between light and glucose regulation controls toxin production and morphogenesis in *Aspergillus nidulans*. Fungal Genet Biol 47:962–972

Axelrod DE, Gealt M, Pastushok M (1973) Gene control of developmental competence in *Aspergillus nidulans*. Dev Biol 34:9–15

Bahn YS, Xue CY, Idnurm A, Rutherford JC, Heitman J, Cardenas ME (2007) Sensing the environment: lessons from fungi. Nat Rev Microbiol 5:57–69

Bayram O, Braus GH (2012) Coordination of secondary metabolism and development in fungi: the *velvet* family of regulatory proteins. FEMS Microbiol Rev 36:1–24

Bayram O, Krappmann S, Ni M, Bok JW, Helmstaedt K, Valerius O, Braus-Stromeyer S, Kwon N-J, Keller NP, Yu J-H et al (2008) VelB/VeA/LaeA complex coordinates light signal with fungal development and secondary metabolism. Science 320:1504–1506

Bayram O, Braus GH, Fischer R, Rodriguez-Romero J (2010) Spotlight on *Aspergillus nidulans* photosensory systems. Fungal Genet Biol 47:900–908

Bayram O, Bayram OS, Ahmed YL, Maruyama J, Valerius O, Rizzoli SO, Ficner R, Irniger S, Braus GH (2012) The *Aspergillus nidulans* MAPK module AnSte11-Ste50-Ste7-Fus3 controls development and secondary metabolism. PLoS Genet 8:e1002816

Bayram O, Feussner K, Dumkow M, Herrfurth C, Feussner I, Braus GH (2016) Changes of global gene expression and secondary metabolite accumulation during light-dependent *Aspergillus nidulans* development. Fungal Genet Biol 87:30–53

Bennett JW, Fernholz FA, Lee LS (1978) Effect of light on aflatoxins, anthraquinones, and sclerotia in *Aspergillus flavus* and *A. parasiticus*. Mycologia 70:104–116

Blumenstein A, Vienken K, Tasler R, Purschwitz J, Veith D, Frankenberg-Dinkel N, Fischer R (2005) The *Aspergillus nidulans* phytochrome FphA represses sexual development in red light. Curr Biol 15:1833–1838

Boylan MT, Mirabito PM, Willett CE, Zimmerman CR, Timberlake WE (1987) Isolation and physical characterization of three essential conidiation genes from *Aspergillus nidulans*. Mol Cell Biol 7:3113–3118

Braus GH, Irniger S, Bayram O (2010) Fungal development and the COP9 signalosome. Curr Opin Microbiol 13:672–676

Brodhun F, Feussner I (2011) Oxylipins in fungi. FEBS J 278:1047–1063

Busby TM, Miller KY, Miller BL (1996) Suppression and enhancement of the *Aspergillus nidulans medusa* mutation by altered dosage of the bristle and stunted genes. Genetics 143:155–163

Busch S, Eckert SE, Krappmann S, Braus GH (2003) The COP9 signalosome is an essential regulator of development in the filamentous fungus *Aspergillus nidulans*. Mol Microbiol 49:717–730

Busch S, Schwier EU, Nahlik K, Bayram O, Helmstaedt K, Draht OW, Krappmann S, Valerius O, Lipscomb WN, Braus GH (2007) An eight-subunit COP9 signalosome with an intact JAMM motif is required for fungal fruit body formation. Proc Natl Acad Sci U S A 104:8089–8094

Calvo AM, Bok J, Brooks W, Keller NP (2004) *veA* is required for toxin and sclerotial production in *Aspergillus parasiticus*. Appl Environ Microbiol 70:4733–4739

Casselton L, Zolan M (2002) The art and design of genetic screens: filamentous fungi. Nat Rev Genet 3:683–697

Chang YC, Timberlake WE (1993) Identification of *Aspergillus brlA* response elements (BREs) by genetic selection in yeast. Genetics 133:29–38

Chang MH, Chae KS, Han DM, Jahng KY (2004) The GanB Galpha-protein negatively regulates asexual sporulation and plays a positive role in conidial germination in *Aspergillus nidulans*. Genetics 167:1305–1315

Clutterbuck AJ (1969) A mutational analysis of conidial development in *Aspergillus nidulans*. Genetics 63:317–327

d'Enfert C (1997) Fungal spore germination: Insights from the molecular genetics of *Aspergillus nidulans* and *Neurospora crassa*. Fungal Genet Biol 21:163–172

D'Souza CA, Lee BN, Adams TH (2001) Characterization of the role of the FluG protein in asexual development of *Aspergillus nidulans*. Genetics 158:1027–1036

De Souza CP, Hashmi SB, Osmani AH, Andrews P, Ringelberg CS, Dunlap JC, Osmani SA (2013) Functional analysis of the *Aspergillus nidulans* kinome. PLoS One 8:e58008

Duran RM, Cary JW, Calvo AM (2007) Production of cyclopiazonic acid, aflatrem, and aflatoxin by *Aspergillus flavus* is regulated by *veA*, a gene necessary for sclerotial formation. Appl Microbiol Biotechnol 73:1158–1168

Dutton JR, Johns S, Miller BL (1997) StuAp is a sequence-specific transcription factor that regulates developmental complexity in *Aspergillus nidulans*. EMBO J 16:5710–5721

Dyer PS, O'Gorman CM (2011) A fungal sexual revolution: *Aspergillus* and *Penicillium* show the way. Curr Opin Microbiol 14:649–654

Dyer PS, O'Gorman CM (2012) Sexual development and cryptic sexuality in fungi: insights from *Aspergillus* species. FEMS Microbiol Rev 36:165–192

Dyer PS, Paoletti M, Archer DB (2003) Genomics reveals sexual secrets of *Aspergillus*. Microbiology 149:2301–2303

Ebbole DJ (2010) The conidium. In: Cellular and molecular biology of filamentous fungi. ASM Press, Washington, DC, pp 577–590

Etxebeste O, Herrero-Garcia E, Araujo-Bazan L, Rodriguez-Urra AB, Garzia A, Ugalde U, Espeso EA (2009) The bZIP-type transcription factor FlbB regulates distinct morphogenetic stages of colony formation in *Aspergillus nidulans*. Mol Microbiol 73:775–789

Etxebeste O, Garzia A, Espeso EA, Ugalde U (2010) *Aspergillus nidulans* asexual development: making the most of cellular modules. Trends Microbiol 18:569–576

Fisher MC, Henk DA, Briggs CJ, Brownstein JS, Madoff LC, McCraw SL, Gurr SJ (2012) Emerging fungal threats to animal, plant and ecosystem health. Nature 484:186–194

Galagan JE, Calvo SE, Cuomo C, Ma LJ, Wortman JR, Batzoglou S, Lee SI, Basturkmen M, Spevak CC, Clutterbuck J et al (2005) Sequencing of *Aspergillus nidulans* and comparative analysis with *A. fumigatus* and *A. oryzae*. Nature 438:1105–1115

Garzia A, Etxebeste O, Herrero-Garcia E, Fischer R, Espeso EA, Ugalde U (2009) *Aspergillus nidulans* FlbE is an upstream developmental activator of conidiation functionally associated with the putative transcription factor FlbB. Mol Microbiol 71:172–184

Garzia A, Etxebeste O, Herrero-Garcia E, Ugalde U, Espeso EA (2010) The concerted action of bZip and cMyb transcription factors FlbB and FlbD induces *brlA* expression and asexual development in *Aspergillus nidulans*. Mol Microbiol 75:1314–1324

Geiser DM (2009) Sexual structures in *Aspergillus*: morphology, importance and genomics. Med Mycol 47 (Suppl 1):S21–S26

Grahl N, Shepardson KM, Chung D, Cramer RA (2012) Hypoxia and fungal pathogenesis: to air or not to air? Eukaryot Cell 11:560–570

Gugnani HC (2003) Ecology and taxonomy of pathogenic aspergilli. Front Biosci 8:s346–s357

Han KH (2009) Molecular Genetics of *Emericella nidulans* Sexual Development. Mycobiology 37:171–182

Han S, Adams TH (2001) Complex control of the developmental regulatory locus *brlA* in *Aspergillus nidulans*. Mol Gen Genomics 266:260–270

Han DM, Han YJ, Lee YH, Jahng KY, Jahng SH, Chae KS (1990) Inhibitory conditions of asexual development and their application for the screening of mutants defective in sexual development. Kor J Mycol 18:225–232

Han S, Navarro J, Greve RA, Adams TH (1993) Translational repression of *brlA* expression prevents premature development in *Aspergillus*. EMBO J 12:2449–2457

Han DM, Han YJ, Kim JH, Jahng KY, Chung YS, Chung JH, Chae KS (1994) Isolation and characterization of NSD mutants in *Aspergillus nidulans*. Kor J Mycol 22:1–7

Han KH, Cheong SS, Hoe HS, Han DM (1998) Characterization of several NSD mutants of *Aspergillus nidulans* that never undergo sexual development. Kor J Genet 20:257–264

Han KH, Han KY, Yu JH, Chae KS, Jahng KY, Han DM (2001) The *nsdD* gene encodes a putative GATA-type transcription factor necessary for sexual development of *Aspergillus nidulans*. Mol Microbiol 41:299–309

Han KH, Lee DB, Kim JH, Kim MS, Han KY, Kim WS, Park YS, Kim HB, Han DM (2003) Environmental factors affecting development of *Aspergillus nidulans*. J Microbiol 41:34–40

Han KH, Seo JA, Yu JH (2004a) A putative G protein-coupled receptor negatively controls sexual development in *Aspergillus nidulans*. Mol Microbiol 51:1333–1345

Han KH, Seo JA, Yu JH (2004b) Regulators of G-protein signalling in *Aspergillus nidulans*: RgsA downregulates stress response and stimulates asexual sporulation through attenuation of GanB (Galpha) signalling. Mol Microbiol 53:529–540

Harris SD (2006) Cell polarity in filamentous. Int Rev Cytol 251:41–77

Harris SD (2008) Branching of fungal hyphae: regulation, mechanisms and comparison with other branching systems. Mycologia 100:823–832

Harris SD (2011) Hyphal morphogenesis: an evolutionary perspective. Fungal Biol 115:475–484

Hicks JK, Yu J-H, Keller NP, Adams TH (1997) *Aspergillus* sporulation and mycotoxin production both require inactivation of the FadA G alpha protein-dependent signaling pathway. EMBO J 16:4916–4923

Hoffmann B, Wanke C, Lapaglia SK, Braus GH (2000) c-Jun and RACK1 homologues regulate a control point for sexual development in *Aspergillus nidulans*. Mol Microbiol 37:28–41

Ichinomiya M, Ohta A, Horiuchi H (2005) Expression of asexual developmental regulator gene *abaA* is affected in the double mutants of classes I and II chitin synthase genes, *chsC* and *chsA*, of *Aspergillus nidulans*. Curr Genet 48:171–183

Jun SC, Lee SJ, Park HJ, Kang JY, Leem YE, Yang TH, Chang MH, Kim JM, Jang SH, Kim HG et al (2011) The MpkB MAP kinase plays a role in post-karyogamy processes as well as in hyphal anastomosis during sexual development in *Aspergillus nidulans*. J Microbiol 49:418–430

Kafer E (1965) Origins of translocations in *Aspergillus nidulans*. Genetics 52:217–232

Kang JY, Chun J, Jun SC, Han DM, Chae KS, Jahng KY (2013) The MpkB MAP kinase plays a role in autolysis and conidiation of *Aspergillus nidulans*. Fungal Genet Biol 61:42–49

Kawasaki L, Sanchez O, Shiozaki K, Aguirre J (2002) SakA MAP kinase is involved in stress signal transduction, sexual development and spore viability in *Aspergillus nidulans*. Mol Microbiol 45:1153–1163

Keller NP, Turner G, Bennett JW (2005) Fungal secondary metabolism – From biochemistry to genomics. Nat Rev Microbiol 3:937–947

Kim H, Han K, Kim K, Han D, Jahng K, Chae K (2002) The *veA* gene activates sexual development in *Aspergillus nidulans*. Fungal Genet Biol 37:72–80

Kim HR, Chae KS, Han KH, Han DM (2009) The *nsdC* gene encoding a putative C_2H_2-type transcription factor is a key activator of sexual development in *Aspergillus nidulans*. Genetics 182:771–783

Kim YJ, Yu YM, Maeng PJ (2017) Differential control of asexual development and sterigmatocystin biosynthesis by a novel regulator in *Aspergillus nidulans*. Sci Rep 7:46340

Kong Q, Wang L, Liu Z, Kwon N-J, Kim SC, Yu J-H (2013) Gbeta-like CpcB plays a crucial role for growth and development of *Aspergillus nidulans* and *Aspergillus fumigatus*. PLoS One 8:e70355

Kovacs Z, Szarka M, Kovacs S, Boczonadi I, Emri T, Abe K, Pocsi I, Pusztahelyi T (2013) Effect of cell wall integrity stress and RlmA transcription factor on asexual development and autolysis in *Aspergillus nidulans*. Fungal Genet Biol 54:1–14

Krijgsheld P, Bleichrodt R, van Veluw GJ, Wang F, Muller WH, Dijksterhuis J, Wosten HA (2013) Development in *Aspergillus*. Stud Mycol 74:1–29

Kwon N-J, Garzia A, Espeso EA, Ugalde U, Yu J-H (2010a) FlbC is a putative nuclear C_2H_2 transcription factor regulating development in *Aspergillus nidulans*. Mol Microbiol 77:1203–1219

Kwon N-J, Shin K-S, Yu J-H (2010b) Characterization of the developmental regulator FlbE in *Aspergillus fumigatus* and *Aspergillus nidulans*. Fungal Genet Biol 47:981–993

Kwon N-J, Park H-S, Jung S, Kim SC, Yu J-H (2012) The putative guanine nucleotide exchange factor RicA mediates upstream signaling for growth and development in *Aspergillus*. Eukaryot Cell 11:1399–1412

Lafon A, Seo JA, Han KH, Yu J-H, d'Enfert C (2005) The heterotrimeric G-protein GanB(alpha)-SfaD (beta)-GpgA(gamma) is a carbon source sensor involved in early cAMP-dependent germination in *Aspergillus nidulans*. Genetics 171:71–80

Lara-Rojas F, Sanchez O, Kawasaki L, Aguirre J (2011) *Aspergillus nidulans* transcription factor AtfA interacts with the MAPK SakA to regulate general stress responses, development and spore functions. Mol Microbiol 80:436–454

Lee BN, Adams TH (1994) The *Aspergillus nidulans fluG* gene is required for production of an extracellular developmental signal and is related to prokaryotic glutamine synthetase I. Genes Dev 8:641–651

Lee BN, Adams TH (1996) FluG and *flbA* function interdependently to initiate conidiophore development in *Aspergillus nidulans* through *brlA* beta activation. EMBO J 15:299–309

Lee JY, Kim LH, Kim HE, Park JS, Han KH, Han DM (2013) A putative APSES transcription factor is necessary for normal growth and development of *Aspergillus nidulans*. J Microbiol 51:800–806

Lee MK, Kwon NJ, Choi JM, Lee IS, Jung S, Yu JH (2014) NsdD is a key repressor of asexual development in *Aspergillus nidulans*. Genetics 197:159–173

Lee MK, Kwon NJ, Lee IS, Jung S, Kim SC, Yu JH (2016) Negative regulation and developmental competence in *Aspergillus*. Sci Rep 6:28874

Lengeler KB, Davidson RC, D'Souza C, Harashima T, Shen WC, Wang P, Pan X, Waugh M, Heitman J (2000) Signal transduction cascades regulating fungal development and virulence. Microbiol Mol Biol Rev 64:746–785

Lew RR (2011) How does a hypha grow? The biophysics of pressurized growth in fungi. Nat Rev Microbiol 9:509–518

Marshall MA, Timberlake WE (1991) *Aspergillus nidulans wetA* activates spore-specific gene expression. Mol Cell Biol 11:55–62

Martinelli S (1976) Conidiation of *Aspergillus nidulans* in submerged culture. Trans Br Mycol Soc 67 (1):121–128

Mirabito PM, Adams TH, Timberlake WE (1989) Interactions of three sequentially expressed genes control temporal and spatial specificity in *Aspergillus* development. Cell 57:859–868

Mooney JL, Yager LN (1990) Light is required for conidiation in *Aspergillus nidulans*. Genes Dev 4:1473–1482

Morton AG (1961) The induction of sporulation in mould fungi. Proc R Soc London Ser B 153:548–569

Nahlik K, Dumkow M, Bayram O, Helmstaedt K, Busch S, Valerius O, Gerke J, Hoppert M, Schwier E, Opitz L et al (2010) The COP9 signalosome mediates transcriptional and metabolic response to hormones, oxidative stress protection and cell wall rearrangement during fungal development. Mol Microbiol 78:964–979

Nevalainen H, Peterson R (2014) Making recombinant proteins in filamentous fungi – are we expecting too much? Front Microbiol 5:75

Ni M, Yu JH (2007) A novel regulator couples sporogenesis and trehalose biogenesis in *Aspergillus nidulans*. PLoS One 2:e970

Ni M, Gao N, Kwon N-J, Shin K-S, Yu J-H (2010) Regulation of *Aspergillus* Conidiation. In: Cellular and Molecular Biology of Filamentous Fungi, pp 559–576

Noble LM, Andrianopoulos A (2013) Reproductive competence: a recurrent logic module in eukaryotic development. Proc R Soc B Biol Sci 280

Oiartzabal-Arano E, Garzia A, Gorostidi A, Ugalde U, Espeso EA, Etxebeste O (2015) Beyond asexual development: modifications in the gene expression profile caused by the absence of the *Aspergillus nidulans* transcription factor FlbB. Genetics 199:1127–1142

Palmer DA, Thompson JK, Li L, Prat A, Wang P (2006) Gib2, a novel Gbeta-like/RACK1 homolog, functions as a Gbeta subunit in cAMP signaling and is essential in Cryptococcus neoformans. J Biol Chem 281:32596–32605

Paoletti M, Seymour FA, Alcocer MJ, Kaur N, Calvo AM, Archer DB, Dyer PS (2007) Mating type and the genetic basis of self-fertility in the model fungus *Aspergillus nidulans*. Curr Biol 17:1384–1389

Park HS, Yu JH (2012) Genetic control of asexual sporulation in filamentous fungi. Curr Opin Microbiol 15:669–677

Park BC, Park YH, Park HM (2003) Activation of *chsC* transcription by AbaA during asexual development of *Aspergillus nidulans*. FEMS Microbiol Lett 220:241–246

Park HS, Ni M, Jeong KC, Kim YH, Yu JH (2012) The role, interaction and regulation of the velvet regulator VelB in *Aspergillus nidulans*. PLoS One 7: e45935

Park H-S, Nam TY, Han KH, Kim SC, Yu J-H (2014) VelC positively controls sexual development in *Aspergillus nidulans*. PLoS One 9:e89883

Park HS, Yu YM, Lee MK, Maeng PJ, Kim SC, Yu JH (2015) Velvet-mediated repression of beta-glucan synthesis in *Aspergillus nidulans* spores. Sci Rep 5:10199

Park HS, Jun SC, Han KH, Hong SB, Yu JH (2017) Diversity, Application, and Synthetic Biology of Industrially Important *Aspergillus* Fungi. Adv Appl Microbiol 100:161–202

Prade RA, Timberlake WE (1993) The *Aspergillus nidulans brlA* regulatory locus consists of overlapping transcription units that are individually required for conidiophore development. EMBO J 12:2439–2447

Purschwitz J, Muller S, Kastner C, Schoser M, Haas H, Espeso EA, Atoui A, Calvo AM, Fischer R (2008) Functional and physical interaction of blue- and red-light sensors in *Aspergillus nidulans*. Curr Biol 18:255–259

Purschwitz J, Muller S, Fischer R (2009) Mapping the interaction sites of Aspergillus nidulans phytochrome FphA with the global regulator VeA and the White Collar protein LreB. Mol Gen Genomics 281:35–42

Rai JN, Tewari JP, Sinha AK (1967) Effect of environmental conditions on sclerotia and cleistothecia production in *Aspergillus*. Mycopathol Mycol Appl 31:209–224

Ramamoorthy V, Dhingra S, Kincaid A, Shantappa S, Feng X, Calvo AM (2013) The putative C_2H_2 transcription factor MtfA is a novel regulator of secondary metabolism and morphogenesis in *Aspergillus nidulans*. PLoS One 8:e74122

Rauscher S, Pacher S, Hedtke M, Kniemeyer O, Fischer R (2016) A phosphorylation code of the *Aspergillus nidulans* global regulator VelvetA (VeA) determines specific functions. Mol Microbiol 99:909–924

Riquelme M (2013) Tip Growth in Filamentous Fungi: A Road Trip to the Apex. Annual Rev Microbiol 67:587–609

Rodriguez-Romero J, Hedtke M, Kastner C, Muller S, Fischer R (2010) Fungi, hidden in soil or up in the air: light makes a difference. Annu Rev Microbiol 64:585–610

Rodriguez-Urra AB, Jimenez C, Nieto MI, Rodriguez J, Hayashi H, Ugalde U (2012) Signaling the induction of sporulation involves the interaction of two secondary metabolites in *Aspergillus nidulans*. ACS Chem Biol 7:599–606

Rohrig J, Yu Z, Chae KS, Kim JH, Han KH, Fischer R (2017) The *Aspergillus nidulans* Velvet-interacting protein, VipA, is involved in light-stimulated heme biosynthesis. Mol Microbiol 105:825–838

Rosen S, Yu J-H, Adams TH (1999) The *Aspergillus nidulans sfaD* gene encodes a G protein beta subunit that is required for normal growth and repression of sporulation. EMBO J 18:5592–5600

Ruger-Herreros C, Rodriguez-Romero J, Fernandez-Barranco R, Olmedo M, Fischer R, Corrochano LM, Canovas D (2011) Regulation of conidiation by light in *Aspergillus nidulans*. Genetics 188:809–822

Samson RA, Visagie CM, Houbraken J, Hong SB, Hubka V, Klaassen CH, Perrone G, Seifert KA, Susca A, Tanney JB et al (2014) Phylogeny, identification and nomenclature of the genus *Aspergillus*. Stud Mycol 78:141–173

Sarikaya Bayram O, Bayram O, Valerius O, Park HS, Irniger S, Gerke J, Ni M, Han KH, Yu JH, Braus GH (2010) LaeA control of velvet family regulatory proteins for light-dependent development and fungal cell-type specificity. PLoS Genet 6:e1001226

Saxena RK, Sinha U (1973) Conidiation of *Aspergillus nidulans* in submerged liquid culture. Gen Appl Microbiol 19:141–146

Scherer M, Fischer R (1998) Purification and characterization of laccase II of *Aspergillus nidulans*. Arch Microbiol 170:78–84

Schoustra S, Rundle HD, Dali R, Kassen R (2010) Fitness-associated sexual reproduction in a filamentous fungus. Curr Biol 20:1350–1355

Seo JA, Guan Y, Yu JH (2003) Suppressor mutations bypass the requirement of *fluG* for asexual sporulation and sterigmatocystin production in *Aspergillus nidulans*. Genetics 165:1083–1093

Seo JA, Han KH, Yu JH (2004) The *gprA* and *gprB* genes encode putative G protein-coupled receptors required for self-fertilization in *Aspergillus nidulans*. Mol Microbiol 53:1611–1623

Seo JA, Han KH, Yu JH (2005) Multiple roles of a heterotrimeric G-protein gamma-subunit in governing growth and development of *Aspergillus nidulans*. Genetics 171:81–89

Seo JA, Guan Y, Yu JH (2006) FluG-dependent asexual development in *Aspergillus nidulans* occurs via derepression. Genetics 172:1535–1544

Sewall TC, Mims CW, Timberlake WE (1990a) *abaA* controls phialide differentiation in *Aspergillus nidulans*. Plant Cell 2:731–739

Sewall TC, Mims CW, Timberlake WE (1990b) Conidium differentiation in *Aspergillus nidulans* wild-type and *wet-white* (*wetA*) mutant strains. Dev Biol 138:499–508

Shimizu K, Keller NP (2001) Genetic involvement of a cAMP-dependent protein kinase in a G protein signaling pathway regulating morphological and chemical transitions in *Aspergillus nidulans*. Genetics 157:591–600

Si H, Rittenour WR, Xu K, Nicksarlian M, Calvo AM, Harris SD (2012) Morphogenetic and developmental functions of the *Aspergillus nidulans* homologues of the yeast bud site selection proteins Bud4 and Axl2. Mol Microbiol 85:252–270

Skromne I, Sanchez O, Aguirre J (1995) Starvation stress modulates the expression of the *Aspergillus nidulans brlA* regulatory gene. Microbiology 141:21–28

Sohn KT, Yoon KS (2002) Ultrastructural Study on the Cleistothecium Development in *Aspergillus nidulans*. Mycobiology 30:117–127

Song MH, Nah JY, Han YS, Han DM, Chae KS (2001) Promotion of conidial head formation in *Aspergillus oryzae* by a salt. Biotechnol Lett 23:689–691

Steinberg G (2007) Hyphal growth: a tale of motors, lipids, and the Spitzenkorper. Eukaryot Cell 6:351–360

Stinnett SM, Espeso EA, Cobeno L, Araujo-Bazan L, Calvo AM (2007) *Aspergillus nidulans* VeA subcellular localization is dependent on the importin alpha carrier and on light. Mol Microbiol 63:242–255

Timberlake WE (1990) Molecular genetics of *Aspergillus* development. Annu Rev Genet 24:5–36

Tisch D, Schmoll M (2010) Light regulation of metabolic pathways in fungi. Appl Microbiol Biotechnol 85:1259–1277

Todd RB, Davis MA, Hynes MJ (2007) Genetic manipulation of *Aspergillus nidulans*: heterokaryons and diploids for dominance, complementation and haploidization analyses. Nat Protoc 2:822–830

Treseder KK, Lennonb JT (2015) Fungal Traits That Drive Ecosystem Dynamics on Land. Microbiol Mol Biol R 79:243–262

Tsitsigiannis DI, Keller NP (2007) Oxylipins as developmental and host-fungal communication signals. Trends Microbiol 15:109–118

Tsitsigiannis DI, Zarnowski R, Keller NP (2004) The lipid body protein, PpoA, coordinates sexual and asexual sporulation in *Aspergillus nidulans*. J Biol Chem 279:11344–11353

Tsitsigiannis DI, Kowieski TM, Zarnowski R, Keller NP (2005) Three putative oxylipin biosynthetic genes integrate sexual and asexual development in *Aspergillus nidulans*. Microbiology 151:1809–1821

Vallim MA, Miller KY, Miller BL (2000) *Aspergillus* SteA (sterile12-like) is a homeodomain-C_2/H_2-Zn^{2+} finger transcription factor required for sexual reproduction. Mol Microbiol 36:290–301

van Burik JAH, Magee PT (2001) Aspects of fungal pathogenesis in humans. Annu Rev Microbiol 55:743–772

Vienken K, Fischer R (2006) The $Zn(II)_2Cys_6$ putative transcription factor NosA controls fruiting body formation in *Aspergillus nidulans*. Mol Microbiol 61:544–554

Vienken K, Scherer M, Fischer R (2005) The Zn $(II)_2Cys_6$ putative *Aspergillus nidulans* transcription factor repressor of sexual development inhibits sexual development under low-carbon conditions and in submersed culture. Genetics 169:619–630

Wei H, Requena N, Fischer R (2003) The MAPKK kinase SteC regulates conidiophore morphology and is essential for heterokaryon formation and sexual development in the homothallic fungus *Aspergillus nidulans*. Mol Microbiol 47:1577–1588

Wieser J, Lee BN, Fondon J III, Adams TH (1994) Genetic requirements for initiating asexual development in *Aspergillus nidulans*. Curr Genet 27:62–69

Wieser J, Yu J-H, Adams TH (1997) Dominant mutations affecting both sporulation and sterigmatocystin biosynthesis in *Aspergillus nidulans*. Curr Genet 32:218–224

Wong KH, Hynes MJ, Todd RB, Davis MA (2009) Deletion and overexpression of the *Aspergillus nidulans* GATA factor AreB reveals unexpected pleiotropy. Microbiology 155:3868–3880

Wu J, Miller BL (1997) *Aspergillus* asexual reproduction and sexual reproduction are differentially affected by transcriptional and translational mechanisms regulating stunted gene expression. Mol Cell Biol 17:6191–6201

Xu JR (2000) Map kinases in fungal pathogens. Fungal Genet Biol 31:137–152

Yager LN, Kurtz MB, Champe SP (1982) Temperature-shift analysis of conidial development in *Aspergillus nidulans*. Dev Biol 93:92–103

Yu J-H (2006) Heterotrimeric G protein signaling and RGSs in *Aspergillus nidulans*. J Microbiol 44:145–154

Yu J-H (2010) Regulation of Development in *Aspergillus nidulans* and *Aspergillus fumigatus*. Mycobiology 38:229–237

Yu JH, Wieser J, Adams TH (1996) The *Aspergillus* FlbA RGS domain protein antagonizes G protein signaling to block proliferation and allow development. EMBO J 15:5184–5190

Zonneveld BJM (1988) Effect of carbon dioxide on fruiting in *Aspergillus nidulans*. Trans Br Mycol Soc 91:625

Signals and Development

Biomechanics of Hyphal Growth

Roger R. Lew[1]

CONTENTS

I. Introduction (Why Hyphal Growth?)

Fungi are a complex group of organisms, comprised of at least three related phylogenetic clades. They are—for the most part—saprobes but can also be pathogens and symbiotic partners. As saprobes, their ability to decompose organic material is crucial to global carbon cycling. They have a characteristic growth pattern—which is fundamentally pleiomorphic in nature. Leading hyphae extend into new territory to find organic material to fuel continued growth. Much of the digestion of organic material in the invaded territory occurs through the action of biodegrading enzymes secreted from the fungal cells. Behind the leading hyphae, a mycelial network of considerable complexity develops, ramifying through the substrate and taking up nutrients that have been decomposed through the action of the biodegrading enzymes. These nutrients in turn provide the metabolic energy necessary for continued growth at the hyphal colony edge. In practically all circumstances, growth requires that the hyphae be able to physically penetrate the substrate. Penetration is also a growth characteristic of pathogens and symbionts, since in both cases, the fungi must "invade" the host tissue and/or cells. The pleiomorphic nature of growth patterning arises from the ability of the fungi to adapt and grow in new locations dependent on local nutrient supply and obstacles to growth.

Like most bacteria and plants, fungi have an extracellular wall that provides a protective structure surrounding the cytoplasm. Because the concentrations of various solutes within the hyphal cell are usually higher than that in the external environment, water is taken up, creating a hydrostatic pressure—turgor—that puts the outer wall under considerable tensile stress. With a few exceptions that will be discussed later, turgor is the major driving force for cell expansion and the penetration of substrates by the fungus.

In this chapter, recent advances in our understanding of the biomechanics of hyphal growth will be presented.

II. Physical Description of the Hyphae

A. Biomechanics

One analogy for pressurized fungal cells are tires of bicycles and cars. When pressure expands the tire, it rigidifies the outer structure of the tire. The outer structure is often a composite that includes wire to withstand the ten-

[1] Department of Biology, York University, Toronto, Canada; e-mail: planters@yorku.ca

Biology of the Fungal Cell, 3rd Edition
The Mycota VIII
D. Hoffmeister, M. Gressler (Eds.)

sile stress caused by the internal pressure; the wire is embedded in an elastic material, rubber. The fungal wall similarly provides elements to provide tensile strength and elastic flexibility. The composite wall structure is far more complex than a tire, comprised of interacting polymeric molecules and can be considered a nanostructural bio-composite material.

For pressurized structures like hyphae, the mechanics of hyphal strength are shown in Fig. 1. Cell wall mechanical strength is explained clearly by Nobel (1991). When the cell is under pressure, the outer structural wall experiences **tension** along its surface. The magnitude of the tension varies dependent on the geometry and direction of the tensional stress. For a sphere, the **tangential stress** (σ_S) (in units of Pascals) is described by $\sigma_S = rP/2t$, where r is the **radius** of the sphere, P is the **pressure**, and t is the **thickness** of the wall. Basically, tensional stress increases with either increasing size or pressure and decreases with increasing wall thickness. The same tension is experienced by cylindrical structures in the longitudinal direction (the direction of cylindrical elongation) ($\sigma_L = rP/2t$), but not tangentially (i.e., radial expansion of the cylindrical cell): $\sigma_T = rP/t$. The latter tangential stress is sometimes referred to as **circumferential** or **hoop stress.** For a simplified geometry of a hyphal tip (a cylinder with a hemisphere at the tip), the tensile stress is unequal at the juncture between the cylinder and the hemispherical tip. In fact, the juncture does appear to be a common location for cell rupture under high pressure (experimentally observed in root hairs of higher plants; Lew 1996).

The tensional stresses on the wall have to be counteracted by the wall. The strength of the wall is related directly to the cell wall material and its orientation within the cell wall. Because of the complex polymer-based nature of the structural architecture, bio-composites do not lend themselves to simple engineering analysis.

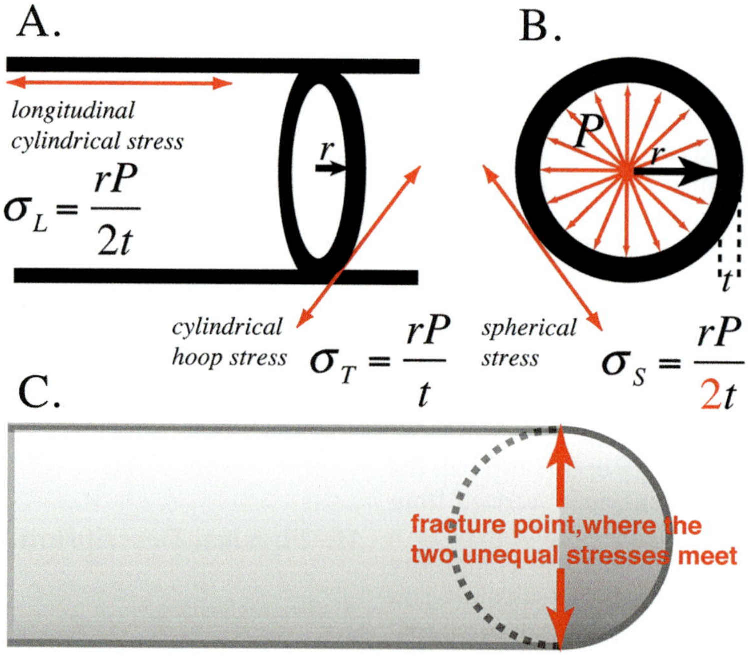

Fig. 1 Hyphal stress mechanics. The tensile stresses on the hypha vary dependent on the geometry. (**a**) The cylindrical geometry has two stresses: longitudinal along the axis of the hyphal cylinder ($\sigma_L = rP/2t$, r is the radius, P is the pressure, and t is the wall thickness) and tangential (often called the hoop stress) ($\sigma_T = rP/t$). (**b**) The hemispherical stress ($\sigma_S = rP/2t$) is calculated similarly to the longitudinal stress. (**c**) Where the cylinder and hemisphere meet, there is a mismatch between the cylindrical hoop stress and the hemispherical stress. Experimentally, failure of the wall is observed in this region of mismatched stresses

Niklas and Spatz (2012) describe the complexity of the strength of bio-composites in detail. For a hypha, there are three biomechanical responses to stress that need to be considered: elasticity, stiffness (or rigidity), and plasticity (Fig. 2).

As pressure is applied internally, the wall can respond elastically, similar to a rubber band: expanding and contracting freely as the pressure is varied. But there is a limit beyond which elasticity fails as the pressure is increased because the stretching "thins" the wall, increas-

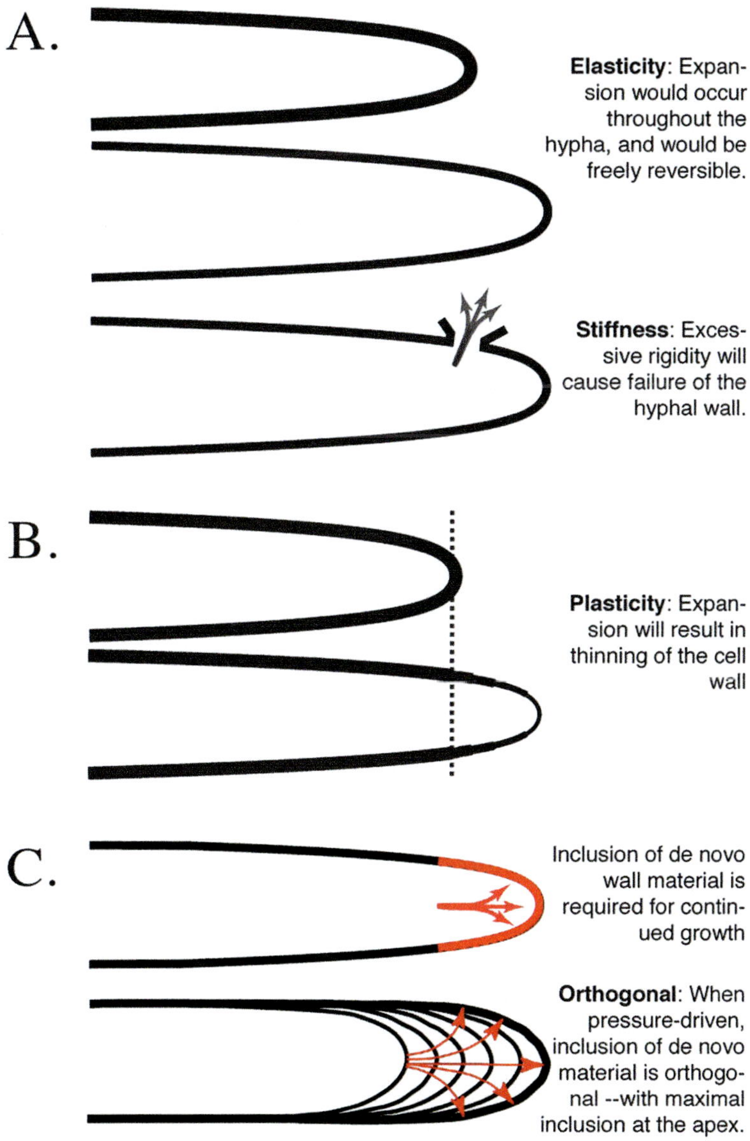

Fig. 2 Model of hyphal expansion. (**a**) The responses to hyphal expansion will depend on the elasticity/stiffness of the hypha. Completely elastic allows for expansion that would be freely reversible. Stiffness (and the hypha will become more stiff as it expands) eventually would result in rupture that would cause cytoplasmic leakage from the cell. (**b**) Plasticity would allow expansion that is not reversible. With increasing hyphal area, the wall material would thin (resulting in vulnerability to rupture). (**c**) Thus, plasticity must occur in tandem with the inclusion of new wall material at the expanding tip. This process is described as orthogonal, such that the inclusion of new material is "spread" through the expanding tip, consistent with the idea that pressure is the driving force for hyphal expansion

ing the tensional stress. At higher pressure, cell walls do tend to lose their ability to respond elastically. So, either wall "thinning" or increased rigidity, or both, eventually causes the cell wall to rupture (Fig. 2a).

Plasticity refers to the ability of the material to change its shape. Ideally, this is not reversible. But biological materials are often viscoelastic, because the deformation is partially reversible if the stress is alleviated. **Viscoelastic deformation** is the common model for cellular expansion in walled cells; it provides an accurate framework to explore the nature of spatially localized expansion in cells across the plant and fungal kingdoms. Viscoelastic responses can vary with the magnitude of the turgor, formalized by Lockhart (1965), who proposed that irreversible cell wall expansion is described by $r\Phi(P-P_E)$, where r is the radius, Φ is the **extensibility** of the wall, P is the **turgor pressure**, and P_E is the **threshold pressure** required before expansion will occur. Above the threshold pressure, the wall becomes plastic and deformable. If the wall composition can be modified to be more viscoelastic at specific locations, then that is where expansion will occur. For hyphal growth, the hyphal tip is the site of expansion, and therefore the tip must be viscoelastic. Direct experimental evidence in support of regulated viscoelastic properties is not easy to come by. The most compelling was obtained from work on the alga *Chara corallina* (Proseus and Boyer 2006, 2007), which is experimentally accessible because of its large size. In *Chara*, the plasticity (or extensibility of the elongating cylinder) is regulated by calcium cross-linkage of pectin in the wall. The more cross-linkages there are, the lower the plasticity of the wall. The nature of regulation of the cell wall plasticity in fungi must be different, since pectin does not contribute to fungal cell wall composition (Free 2013). Instead, remodeling of the wall appears to rely upon the activity of various hydrolases (Adams 2004) that are spatiotemporally regulated.

Viscoelasticity alone is insufficient to allow for continued cellular expansion because the wall thins as its surface area increases (Fig. 2b). Thus, new membrane and wall material must be incorporated as the tip extends (Fig. 2c). Experimental measurements show that hyphal tip expansion is orthogonal (Bartnicki-Garcia et al. 2000). That is, incorporation of new material occurs perpendicular to the hyphal tip surface as the tip extends, resulting in an orthogonal displacement of particles on the cell surface. Bartnicki-Garcia et al. (2000) argue that orthogonal growth must be pressure-driven, since only pressure would apply a force that is always perpendicular to the inside surface of the hyphae. Other experimental evidence certainly supports the central role of turgor in growth.

The elasticity/stiffness/plasticity of the hyphae is described experimentally by a plot of applied force *versus* deformation (or a stress *versus* strain relation in the terminology of engineering). Niklas and Spatz (2012) explain the stress-strain relations of biological materials. Basically, the stress is the force applied to the cell (pressure in our case), while the strain is the deformation of the cell caused by the stress. For an elastic material, the relation would be linear: apply a tensional stress, the material will stretch, decrease the tension and the material will return to its original shape. A rigid or stiff material won't undergo deformation. And plastic deformation is irreversible. Generally, the slope of the stress-strain relation (called the modulus of elasticity) is used to quantitatively describe how much deformation occurs. A shallow slope represents a biomaterial that is easily deformed; a steeper slope indicates a stiff or rigid biomaterial. An example of stress-strain relations for turgid hyphae is presented in Sect. III.

B. Turgor Measurements

There are a number of techniques that can be used to measure the turgor pressure of a walled cell. Some are indirect—such as measuring the osmolarity that causes incipient plasmolysis or **ball tonometry**. One is very direct—impalement of the cell with a pressure probe. The pressure probe technique (Zimmerman et al. 1969) has been used in a wide range of cells (plant, algal, and fungal) for many decades. Hüsken et al. (1978) provide a detailed description of the technique.

The value of direct pressure measurements is clear. The pressure can be continuously monitored during responses to various treatments. In the hands of an experienced cell biophysicist, the cell seldom exhibits any sign of damage and in fact can continue to grow after impalement with the pressure probe (Lew 2005). Upon impalement with a silicon oil-filled micropipette, turgor pushes the oil/cytoplasm meniscus into the micropipette. Pressure is applied to push the meniscus back to the micropipette tip. The applied pressure is an estimate of the initial turgor. For fungal hyphae, some of the more biophysical measurements, such as hydraulic conductivity, may or may not be measurable, because the cell "unit" may not be finite (instead, it is a continual tube of indeterminate length).

An alternative method of measuring turgor is an indirect one: ball tonometry (Lintilhac et al. 2000). In this technique, a ball of known (and small) size is pressed against the surface of the cell being measured. The applied force and the indentation contact area are used to estimate the turgor. The technique is limited: Only surface cells with elastic cell wall responses are measurable.

Despite the limitations, ball tonometry does offer a dynamic way to assess responses to forces coming from the outside of the cell. For example, how much force causes a growth response? Can the applied forces induce a shift to a penetration growth pattern? These and other questions may become directly answerable using ball tonometry.

C. Osmotic Pressure

The turgor pressure of a walled cell arises from the differences in the osmolarity inside and outside the cell, often referred to as the **osmotic pressure**. This is described by the equation: $\Delta P = RT(c_i - c_o)$, in which ΔP is the **pressure**, R is the **gas constant** (8.314 L kPa K^{-1} mol^{-1}), T is the **temperature** (K), and c_i and c_o are the **concentrations** (mol L^{-1}) of **osmotically active solutes** inside and outside the cell, respectively. For a normal turgor of 600 kPa at room temperature (293 K), the cytoplasmic concentration of osmotically active solutes would have to be about 250 mM higher than the external concentration. During growth, the cell volume increases due to water influx, so that the osmotically active solutes inside the cell would become diluted, decreasing turgor. To maintain turgor during expansive growth, osmotically active solutes must be taken up and/or synthesized de novo.

III. Turgor and Growth

Does turgor "drive" growth? Apparently it does. When turgor is adjusted by increasing the external concentration of osmolytes, the turgor (due to the *difference* in osmolyte concentrations inside and outside of the cell) declines, as does growth. An example of the relation between turgor and growth (and subsequent turgor regulation) is shown in Fig. 3 for the fungus *Neurospora crassa* (Lew and Nasserifar 2009). After addition of external osmoticum, the hyphal turgor declines precipitously (Fig. 3a); growth rates drop at the same time (Fig. 3b). The decline in turgor also results in a decrease in hyphal volume (Fig. 3c). Micrographs of a hypha impaled with a pressure probe show the hyphal volume shrinkage at the times specified (Fig. 3d). Over time, turgor, growth, and hyphal volume increase as a consequence of turgor regulation mechanisms to be described in a later section.

The biomechanical behavior of the hyphae can be inferred from the relation between turgor (*stress*) and hyphal shrinkage (*strain*) (Fig. 3e). The slopes of the curve (the modulus of elasticity) provide a measure of the extent of strain that occurs. In general, at low turgor, the hyphae are readily deformed; thus the modulus of elasticity is fairly low (about 0.4 MPa). As turgor is increased, the hyphal wall expands and becomes stiffer, resulting in a higher modulus (about 1.8 MPa). The full range of the modulus is not being measured (i.e., at turgor higher than normal), but the modulus would be expected to increase even more as the wall was stretched to its limit. For walled plants and algal cells, moduli vary from about 1 (very elastic) to 60 MPa (very stiff) (Zimmerman 1978).

The basic observation is that there is a clear relation between turgor and hyphal growth. And in response to hyperosmotic conditions, the turgor is regulated, returning to its initial value (prior to the hyperosmotic treatment) in about 60 min.

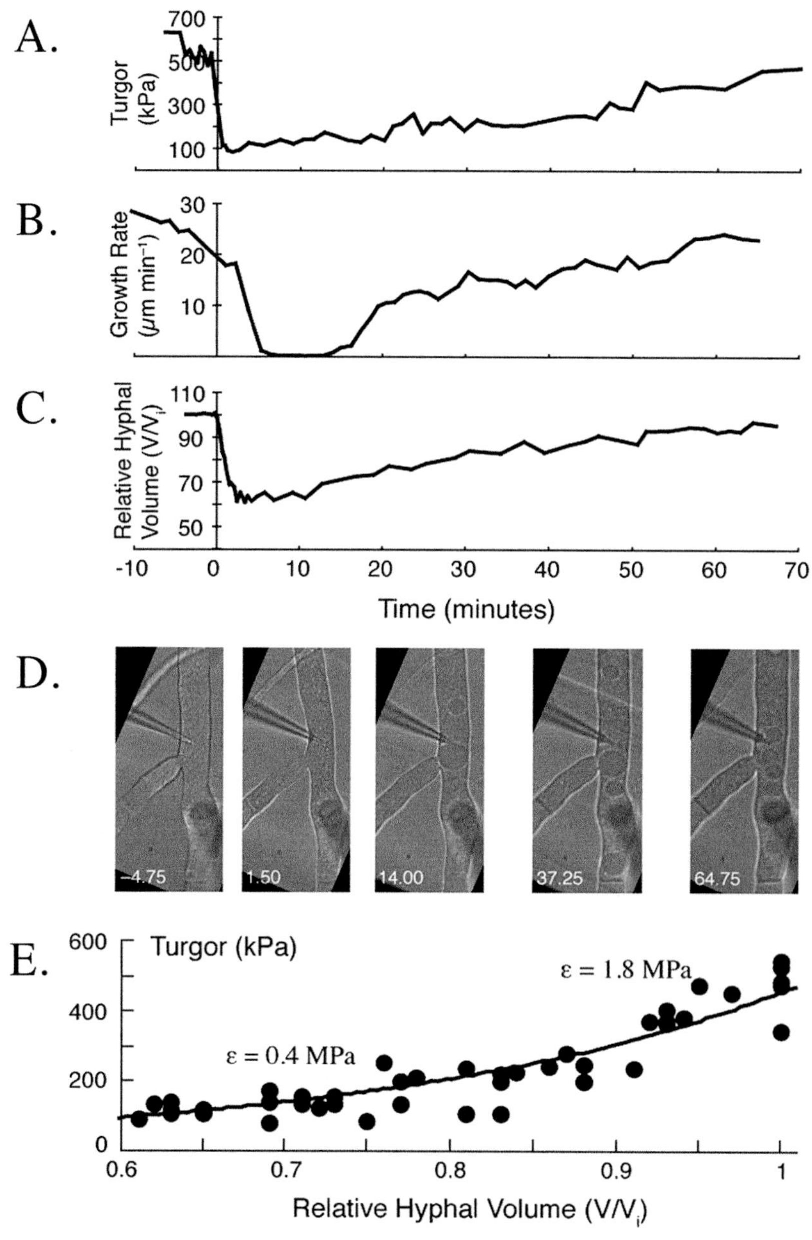

Fig. 3 Turgor regulation of fungal hypha. (**a**) After the addition of a hyperosmotic solution extracellularly, turgor rapidly decreases, but then recovers over 70 min. (**b**) Growth rates respond similarly to turgor. (**c**) Changes in hyphal volume in response to the hyperosmotic treatment respond similarly to both turgor and growth rates. (**d**) Hyphal volume is estimated by measuring hyphal diameters at the site where turgor is being measured with the pressure probe. (**e**) Plotting the stress (turgor) versus strain (hyphal volume) provides a stress-strain relation that identifies the changes in the modulus of elasticity (ε) for the fungal cell

It is not always the case that there is such a straightforward relation between turgor and growth. The Oomycetes are one example of a hyphal organism that does not appear to regulate turgor (Lew et al. 2004). Kaminskyj et al. (1992) reported that the oomycete *Saprolegnia ferax* does require turgor to grow, but growth rates are independent of turgor over a range of about 100–300 kPa. Harold et al. (1996) reported that this hyphal organism was able to grow even in the absence of any measurable turgor. What is difficult to determine is whether growth independent of turgor is due to changes in the threshold pressure required for growth, meaning that the viscoelastic properties of the wall are increasing at lower turgor. There is an alternative mechanism for growth that may explain hyphal extension in the absence of turgor—apical extension caused by the cytoskeleton (Heath and Steinberg 1999). An unusual example of growth in the absence of turgor is the *slime* mutant of *Neurospora crassa* (Emerson 1963; Perkins et al. 1982): This is a triple mutant (*fz; sg; os-1*) that is wall-less, osmosensitive, and exhibits a phenotype of ameboidal growth, with pseudopodial extensions. A role for the cytoskeleton in tip growth is indicated by analysis of cytoskeleton architecture in hyphae penetrating hard substrates (Walker et al. 2006). Certainly, the cytoskeleton does play a role in cytological organization within the growing hyphae (Lew 2011).

IV. Turgor Regulation

To begin with, we need to distinguish between **osmotolerant growth** and **turgor regulation.** Fungal growth at high external osmolarity is physiologically complex (Jennings 1995). Generally, fungi are able to grow at elevated NaCl concentrations of 1.5–5 M (Griffin 1994) or even saturated salt solutions for extremophiles (Gostinčar et al. 2009). At such high external concentrations of osmotically active solutes, turgor has to decline (the cell is unlikely to have a cytoplasmic salt concentration higher than the saturated solution outside).

For non-extremophiles, the response to high extracellular osmoticum is complex. First, if the wall is elastic, the cell volume will shrink causing the internal osmolarity to increase. This can offset—to varying degrees—the higher external osmolarity. Second, if the cell takes up the external solute, this too will offset a decline in turgor. Finally, changes in the yield threshold of the wall may allow continued growth at lower turgor. Thus, the ability to grow at high external osmolarity may not depend solely upon active turgor regulation. Whether turgor itself is regulated has to be determined by its direct measurement.

There are a number of mutants that are osmosensitive, isolated on the basis of their inability to grow in NaCl (4% NaCl is often used); the causes of their osmosensitivity vary (Radford 2014). The best-known *Neurospora crassa* mutants are the *os* (osmotic sensitive) mutants. Many of the *os* genes encode for components of a signal transduction cascade that induce multiple responses to high external osmotic shock and do play a role in turgor regulation (described later). There are other osmotic sensitive mutants, such as *eas* (easily wettable) and *sor-4* (insensitive to growth restriction caused by sorbose). The wild-type *eas* gene encodes for hydrophobin, which creates a non-wettable, hydrophobic layer on conidia. The wild-type *sor-4* encodes for a sugar sensor. So, osmosensitivity has many different underlying causes. I will focus on turgor and its regulation.

Direct evidence for relatively rapid turgor recovery of wild type after hyperosmotic treatment is shown in Fig. 3. Turgor levels under normal growing conditions do vary among osmotic sensitive mutants. Normal turgor values for mutants examined so far in *Neurospora crassa* are compiled in Fig. 4a. Two of the mutants (*os-1* and *os-2*) are the osmosensor and final kinase in a **MAP kinase cascade.** When activated by high external osmoticum, they activate changes in ion transport across the plasma membrane and the synthesis of the osmolyte glycerol. Ion uptake and glycerol synthesis both contribute to recovery of turgor to initial levels (Fig. 4b). The dual responses of the signaling cascade operate differently. Increased glycerol synthesis requires changes in gene expression (Klipp et al. 2005) and is relatively slow (about 60 min) (Lew and Levina 2007), while activation of ion uptake is direct (activating both the plasma membrane H^+-ATPase and K^+ uptake) and faster (about 10 min) (Lew et al. 2006). Turgor recovery can rely solely on ion uptake, since the cut mutant—incapable of synthesizing glycerol—regulates its turgor (Lew and Levina 2007). Some of the other mutants that exhibit lower turgor under normal conditions (*mid-1*, *ptk-2*, and *plc-1*) probably play a regulatory role in downstream responses (Fig. 4c).

A.

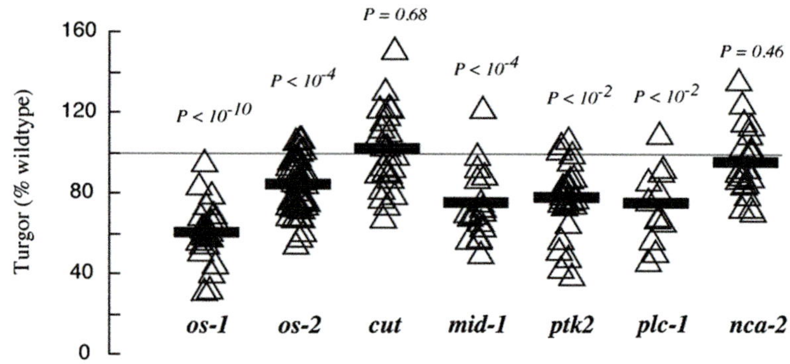

B.

	WT	*os-1*	*os-2*	*cut*	*mid-1*	*ptk2*
H^+-ATPase activation	+	–	–	+	+	+
K^+ Uptake	+	–		+		
Cl^- Uptake	+	+		+		
Glycerol	+	–	–	–		
Turgor Recovery	+	–	+/–	+	+	+

C.

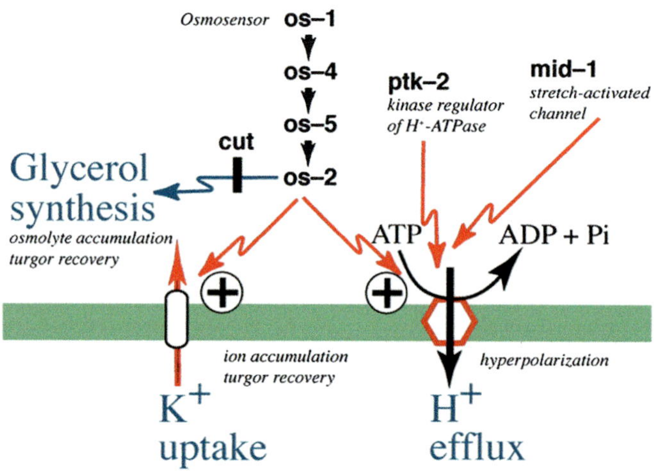

Fig. 4 Genes functioning in turgor regulation. (**a**) The normal turgor values of a number of mutants can be similar to or lower than wild-type turgor. The data are shown as the percent of wild-type turgor measurements that were performed as controls for the experimental runs measuring the turgor of the mutants. Mutants that exhibit lower turgor are *os-1*, *os-2*, *mid-1*, *ptk-2*, and *plc-1*. The *os-1* and *os-2* genes are part of a MAP kinase cascade that functions in osmo-tolerance. The *mid-1* gene encodes for a stretch-activated channel. The *ptk-2* gene encodes for a protein that regulates ion transport activity (the plasma membrane H^+-ATPase). The *plc-1* gene encodes for a phospholipase C. (**b**) For the mutants, various aspects of turgor regulation (H^+-ATPase activation, K^+ and Cl^- uptake, glycerol synthesis, and turgor recovery) have been explored, although much remains to be done. (**c**) A diagrammatic description of the potential roles of the various genes examined so far

V. Cytoplasmic Movement (Mass Flow)

The accumulation of ions and osmolytes maintains sustained hyphal elongation. But in addition, cytoplasm is recruited to the growing edge of the colony by **mass flow** (Lew 2005). The driving force for mass flow is pressure differences along the hyphal tubes.

A. Low Reynolds Number Hydrodynamics

Mass flow in hyphae occurs at low Reynolds number (Reynolds 1883). So, the flow is laminar, there is no turbulence. The Reynolds number is a dimensionless value, calculated from the equation: $Re = (\rho v 2r)/\eta$ where ρ is the **density** (about 1 g cm^{-3}), v is the **velocity of flow** (about 5×10^{-4} cm s^{-1}), $2r$ is the **diameter** of the **hyphal tube**, and η is the **viscosity** (similar to water, about 0.01 g s^{-1} cm^{-1}). The value of Re is the ratio of inertial to viscous forces. At high Re, inertial forces dominate, and flow is turbulent; at low values, viscous forces dominate, so that flow is laminar in nature. For cytoplasm movements in fungi, estimates of the Reynolds number are on the order of 10^{-4}. The Reynolds number at which turbulent flow will occur in a hydraulic pipe is on the order of 1 to 10^3 (Brody et al. 1996).

At low Reynolds number, the flow velocity can be estimated by the Hagen-Poiseuille equation: $v = [(\Delta P_{hyphal} / l)r^2]/(8\eta)$. The pressure gradient ($\Delta P_{hyphal} / l$) required to cause velocities of 5×10^{-4} cm s^{-1} is about 0.05 kPa cm^{-1} (Lew 2005). This is a very small pressure difference compared to the normal turgor of the hypha (400–500 kPa).

The nature of mass flow through the hyphal tubes is complicated by the presence of organelles moving in the cytoplasmic stream. Experimental evidence of the non-ideal nature of flow is based on measurements of the velocity profile within the hypha. Hagen-Poiseuille predicts a parabolic shape, with maximal velocities at the center of the cylinder, tapering to zero at the hyphal walls. The actual profile is flat (Abadeh and Lew 2013), a shape that is consistent with "partial plug flow" (Cox and Mason 1971).

Basically, the organelles moving within the tube (nuclei, mitochondria, and vacuole velocity profiles have been mapped, Abadeh and Lew 2013) modify the pattern of fluid shearing that would normally result in parabolic velocity profiles. Even so, estimates of the pressure gradients using Hagen-Poiseuille will not be that different from the pressure dependence of "partial plug flow" (Cox and Mason 1971).

B. Effect of Hyphal Pores

Do septal pores impede mass flow? The simple answer is: Not by much. The reason is that the pore may be smaller than the hyphal diameter, but the length of the pore is very short, so the pore wall impedes flow over a very small distance, while the walls of the hyphal cylinder create a frictional drag over distances that are considerably longer.

Vogel (2003) provides a clear and straightforward explanation of the effect of pores and their relevance to mass transport in organisms besides fungi. Happel and Brenner (1986) provide a more complete discussion. At low Reynolds number, the physical descriptions of flow through the hyphal tube and through the pore are different. For the first, the volumetric rate of flow (Q, in units of m^3 s^{-1}) is:

$$Q = \frac{\Delta P_{hyphal}}{l} \cdot \frac{\pi \cdot R^1}{8 \cdot \eta} \quad (1)$$

where $\Delta P_{hyphal}/l$ is the **pressure gradient** (Pa m^{-1}), R is the **hyphal radius** (m), and η is the **viscosity** (Pa s). For movement through a pore, the volumetric rate of flow is (Happel and Brenner 1986):

$$Q = \frac{\Delta P_{pore} \cdot r^3}{3 \cdot \eta} \quad (2)$$

where ΔP_{pore} is the **pressure drop** across the pore and r is the **pore radius**. An example of a calculation using values that are known for hyphae is presented in Fig. 5. To assess the pressure drop in the hypha, I assumed a hyphal length of about 200 μm between septa.

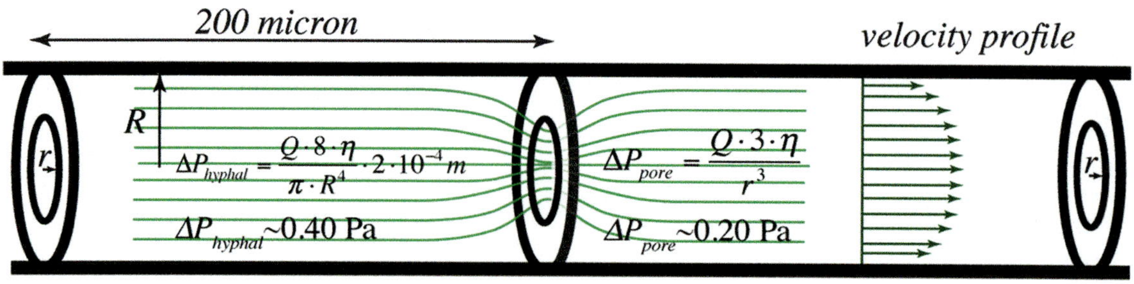

Fig. 5 Micro-fluidics of a hypha. To compare flow through the hyphal cylinder and the narrower aperture of the septal pore, the pressure drops through a 200 micron cylindrical length and through an aperture that was 50% of the hyphal width were calculated for a volume flow of 7.85×10^{-16} m^3 s^{-1} (0.785 pl s^{-1}, the flow that would occur for a cytoplasm velocity of 5 μm s^{-1} in a 20 μm diameter hypha). The pressure drops for flow through the cylinder are similar in magnitude to the pressure drop for flow through the septal pore (0.4 Pa compared to 0.2 Pa). Thus, the pore has some effect—but not a large one—on flow through the hyphal network. Experimentally measured velocity profiles (Abadeh and Lew 2013) are also shown in the diagram

I used a volumetric rate of flow of 7.85×10^{-16} m^3 s^{-1} that was calculated from a typical velocity of particle flow of 5 μm s^{-1}. The pressure drop for mass flow in the 200 μm length of hypha is about 0.40 Pa. With a septal pore radius half of the hyphal diameter, the pressure drop for mass flow through the pore is about 0.2 Pa. Thus, although the septal pore does impede flow, the effect is smaller than might be expected. Johansen (1930) explored a model system that is pertinent to flow through hyphal tubes and septal pores; his results matched the expected result (Happel and Brenner 1986). In his experimental observations of flow lines through the pipe orifice, he found that the flow was completely laminar and symmetrical on either side of the pore at very low Reynolds number (similar to hyphal flow).

Septal pores have relatively small effects on cytoplasmic flow under normal conditions. When there is damage to the mycelial network that causes cytoplasmic loss from the pressurized hyphae, they must seal rapidly to protect the integrity of the cytoplasm. Thus, their formation (Lai et al. 2012) and mechanisms of pore closure are important to the survival of the hyphal organism. Regulation of septal pores also occurs under a variety of non-damaging environmental conditions (van Peer et al. 2009) and even in mitosis (Shen et al. 2014).

VI. Penetration (Invasive Growth)

Another biomechanical aspect of hyphal growth is the ability of the hyphae to penetrate tissues or substrates. How much pressure does a hypha apply to penetrate a substrate? Bastmeyer et al. (2002), Money (2007), and Yafetto et al. (2009) explored aspects of this question for fungi. Indirect techniques and high sensitivity strain gauges can be used to measure the very small forces a small hypha will exert when growing into a substrate. Lab-on-a-chip technologies are being pioneered for other tip-growing organisms as well, such as pollen tubes which must penetrate the stigma to effect successful pollination (Burri et al. 2018).

Adapting growth to penetrate substrates is related to **two sensor-response systems**: chemotropism (Turrà et al. 2016) and thigmotropism (Almeida and Brand 2017). The biochemical mechanisms underlying chemo- and mechano-sensing are very diverse and generally result in activation of MAP kinase and cAMP signaling cascades (Turrà et al. 2016; Braunsdorf et al. 2016). The final result is to initiate penetration into substrate.

Substrate penetration by hyphae—whether into host tissue or ramifying through soil or other substrate—will depend upon physical force. These forces are small on an absolute scale. In *Armillaria* rhizomorphs, Yafetto et al.

(2009) reported values of about 1–6 mN, corresponding to a turgor of 40–300 kPa. Can these pressures be use to mechanically penetrate a tissue or substrate? There is no direct answer. For example, as a hypha penetrates a tissue, the tip will naturally take the path of least resistance. In doing so, it can easily decrease the hyphal tip size to effect easier penetration. Finally, it can modify the structural integrity of the substrate through extracellular hydrolases to soften the substrate. None of these adaptive processes lend themselves to a simple biomechanical analysis. In this regard, recent advances using lab-on-a-chip technologies should be useful. For example, Tayagui et al. (2017) grew hyphae of *Achlya bisexualis* (an oomycete) through a micro-pillar array and measured pillar deflections to quantify the force applied by the hypha. Burri et al. (2018) coupled micro-channels with a microelectromechanical force-sensor to measure the forces at the tips of pollen tubes as they collided with and responded to the force-sensing barrier. In the future, these technical advances should provide methods to enlighten our understanding of fungal penetration of substrates.

References

Abadeh A, Lew RR (2013) Mass flow and velocity profiles in *Neurospora* hyphae: partial plug flow dominates intra-hyphal transport. Microbiology 159: 2386–2394

Adams DJ (2004) Fungal cell wall chitinases and glucanases. Microbiology 150:2029–2035

Almeida MC, Brand AC (2017) Thigmo responses: the fungal sense of touch. Microbiol Spectr 5(2). https://doi.org/10.1128/microbiolspec.FUNK-0040-2016

Bartnicki-Garcia S, Bracker CE, Gierz G, López-Franco R, Lu H (2000) Mapping the growth of fungal hyphae: orthogonal cell wall expansion during tip growth and the role of turgor. Biophys J 79:2382–2390

Bastmeyer M, Deising HB, Bechinger C (2002) Force exertion in fungal infection. Annu Rev Biophys Biomol Struct 31:321–341

Braunsdorf C, Mailänder-Sánchez D, Schaller M (2016) Fungal sensing of host environment. Cell Microbiol 18:1188–1200. https://doi.org/10.1111/cmi.12610

Brody JP, Yager P, Goldstein RE, Austin RH (1996) Biotechnology at low Reynolds numbers. Biophys J 71:3430–3441

Burri JT, Vogler H, Läubli NF, Hu C, Grossniklaus U, Nelson BJ (2018) Feeling the force: how pollen tubes deal with obstacles. https://doi.org/10.1101/266106

Cox RG, Mason SG (1971) Suspended particles in fluid flow through tubes. Annu Rev Fluid Mech 3:291–316. https://doi.org/10.1146/annurev.fl.03.010171.001451

Emerson S (1963) Slime, a plasmodial variant of *Neurospora crassa*. Genetica 34:162–182

Free SJ (2013) Fungal cell wall organization and biosynthesis. Adv Genet 81:33–82

Gostinčar C, Martin Grube M, De Hoog S, Zalar P, Gunde-Cimerman N (2009) Extremotolerance in fungi: evolution on the edge. FEMS Microbiol Ecol 71:2–11

Griffin DH (1994) Fungal physiology. Wiley-Liss, New York, p 144

Happel J, Brenner H (1986) Low Reynolds number hydrodynamics. Martinus Nijhoff, Dordrecht, pp 153–154

Harold RL, Money NP, Harold FM (1996) Growth and morphogenesis in *Saprolegnia ferax*: is turgor required? Protoplasma 191:105–114

Heath IB, Steinberg G (1999) Mechanisms of hyphal tip growth: tube dwelling amebae revisited. Fungal Genet Biol 28:79–93

Hüsken D, Steudle E, Zimmerman U (1978) Pressure probe technique for measuring water relations of cells in higher plants. Plant Physiol 61:158–163

Jennings DH (1995) The physiology of fungal nutrition. Cambridge University Press, Cambridge, pp 398–446

Johansen FC (1930) Flow through pipe orifices at low Reynolds number. Proc R Soc Lond A 126: 231–245. https://doi.org/10.1098/rspa.1930.0004

Kaminskyj SGW, Garrill A, Heath IB (1992) The relation between turgor and tip growth in *Saprolegnia ferax*: turgor is necessary, but not sufficient to explain apical extension rates. Exp Mycol 16:64–75

Klipp E, Nordlander B, Kruger R, Gennemark P, Hoffmann S (2005) Integrative model of the response of yeast to osmotic shock. Nat Biotechnol 23: 975–982

Lai J, Koh CH, Tjota M, Pieuchot L, Raman V, Chandrababu KB, Yang D, Wong L, Jedd G (2012) Intrinsically disordered proteins aggregate at fungal cell-to-cell channels and regulate intercellular connectivity. Proc Natl Acad Sci U S A 109: 15781–15786

Lew RR (1996) Pressure regulation of the electrical properties of growing *Arabidopsis thaliana* L. root hairs. Plant Physiol 112:1089–1100

Lew RR (2005) Mass flow and pressure-driven hyphal extension in *Neurospora crassa*. Microbiology 151: 2685–2692

Lew RR (2011) How does a hypha grow? The biophysics of pressurized growth in fungi. Nat Rev Microbiol 9:509–518

Lew RR, Levina NN (2007) Turgor regulation in the osmosensitive cut mutant of *Neurospora crassa*. Microbiology 153:1530–1537

Lew RR, Levina NN, Shabala L, Anderca MI, Shabala SN (2006) Role of a mitogen-activated protein kinase cascade in ion flux-mediated turgor regulation in fungi. Eukaryot Cell 5:480–487

Lew RR, Levina NN, Walker SK, Garrill A (2004) Turgor regulation in hyphal organisms. Fungal Genet Biol 41:1007–1015

Lew RR, Nasserifar S (2009) Transient responses during hyperosmotic shock in the filamentous fungus *Neurospora crassa*. Microbiology 155:903–911

Lintilhac PM, Wei C, Tanguay JJ, Outwater JO (2000) Ball tonometry: a rapid, nondestructive method for measuring cell turgor pressure in thin-walled plant cells. J Plant Growth Regul 19:90–97. https://doi.org/10.1007/s003440000009

Lockhart JA (1965) An analysis of irreversible plant cell elongation. J Theor Biol 8:264–275

Money NP (2007) Biomechanics of invasive hyphal growth. In: Howard RJ, Gow NAR (eds) Biology of the fungal cell, The Mycota VIII, 2nd edn. Springer, Berlin, pp 237–249

Niklas KJ, Spatz H-C (2012) Plant physics. University of Chicago Press, Chicago

Nobel PS (1991) Physicochemical and environmental plant physiology. Academic Press, San Diego

Perkins DD, Radford A, Newmeyer D, Björkman M (1982) Chromosomal loci of *Neurospora crassa*. Microbiol Rev 46:426–570

Proseus TE, Boyer JS (2006) Calcium pectate chemistry controls growth rate of *Chara corallina*. J Exp Bot 57:3989–4002

Proseus TE, Boyer JS (2007) Tension required for pectate chemistry to control growth in *Chara corallina*. J Exp Bot 58:4283–4292

Radford A (curator) (2014) The Neurospora crassa e-compendium. http://www.bioinf.leeds.ac.uk/~gen6ar/newgenelist/genes/index.html. Accessed 3 Sept 2014

Reynolds O (1883) An experimental investigation of the circumstances which determine whether the motion of water shall be direct or sinuous, and of law of resistance in parallel channels. Philos Trans R Soc Lond 174:935–982

Shen K-F, Osmani AH, Govindaraghavan M, Osmani SA (2014) Mitotic regulation of fungal cell-to-cell connectivity through septal pores involves the NIMA kinase. Mol Biol Cell 25:763–775

Tayagui A, Sun Y, Collings DA, Garrill A, Nock V (2017) An elastomeric micropillar platform for the study of protrusive forces in hyphal invasion. Lab Chip 17:3643. https://doi.org/10.1039/c7lc00725f

Turrà D, Nordzieke D, Vitale S, El Ghalid M, Di Pietro A (2016) Hyphal chemotropism in fungal pathogenicity. Semin Cell Dev Biol 57:69–75. https://doi.org/10.1016/j.semcdb.2016.04.020

van Peer AF, Müller WH, Boekhout T, Lugones LG, Woösten HAB (2009) Cytoplasmic continuity revisited: closure of septa of the filamentous fungus *Schizophyllum commune* in response to environmental conditions. PLoS ONE 4:e5977. https://doi.org/10.1371/journal.pone.000597

Vogel S (2003) Comparative biomechanics. Life's physical world. Princeton University Press, Princeton, NJ, pp 187–203

Walker SK, Chitcholtan K, Yu Y-P, Christenhusz GM, Garrill A (2006) Invasive hyphal growth: an F-actin depleted zone is associated with invasive hyphae of the oomycetes *Achlya bisexualis* and *Phytophthora cinnamomi*. Fungal Genet Biol 43:357–365

Yafetto L, Davis DJ, Money NP (2009) Biomechanics of invasive growth by *Armillaria* rhizomorphs. Fungal Genet Biol 46:688–694

Zimmerman U (1978) Physics of turgor- and osmoregulation. Annu Rev Plant Physiol 29:121–148

Zimmerman U, Räde H, Steudle E (1969) Kontinuierliche druckmessung in pflanzenzellen. Naturwissenschaften 56(12):634

Molecular Signalling During the Ectomycorrhizal Symbiosis

Clément Pellegrin[1], Francis Martin[1], Claire Veneault-Fourrey[1]

CONTENTS

I. Introduction

Microbes are the most abundant forms of life on Earth. Microbial world diversity is remarkable, keeping challenging us after more than 300 years of research on it (Woyke and Rubin 2014). This medley of microorganisms set in a wide range of ecosystems and creates complex interactions within its environment. Among them, thousands of microbial species either bacteria or fungi inhabit the soil, either as organic and inorganic matter decayers or as partners interacting with plant roots. They are key players for soil fertility, consequently sustaining plant productivity.

Around 50,000 fungal species interact with around 250,000 of plant species and form mycorrhiza (van der Heijden et al. 2015). Mycorrhiza is a symbiotic organ made of both fungal hyphae and plant root cells. Fully functional mycorrhizal symbioses allow bidirectional nutrient exchanges between the two partners. Consequences for the plant are an increased ability to overcome abiotic and biotic stresses like drought, polluted or poor soils as well as pathogen attacks (Sikes et al. 2009). Seven types of mycorrhizal symbiosis, varying in structure, host range and ecology are described. The most prevalent mycorrhizal fungi are the arbuscular fungi (AM). They all belong to Glomeromycota, a monophyletic group which appeared with the first land plants, about 400 million years ago colonized herbaceous plants including several crop species (Parniske 2008). Beside AM fungi stands the ectomycorrhizal symbioses which are the most prevalent mycorrhiza occurring in forest ecosystems. The ericoid mycorrhiza is occurring between perennial shrubs species (i.e. blueberry, cranberry) and fungi such as *Rhizoscyphus ericae* or *Oidiodendron maius*. Among other group of mycorrhizal fungi, one can cite the ectendomycorrhiza, monotropoid, orchid and arbutoid mycorrhiza.

The forest ecosystems are of major environmental, economic and scientific interests and are composed of a wide diversity of trees, which dominate boreal, temperate, Mediterranean and tropical environment. This diversity directly participates to forest ecosystems functioning (Isbell et al. 2011), notably by efficient

[1] Laboratoire d'excellence ARBRE, Centre INRA de Nancy-Lorraine, UMR INRA/Université de Lorraine 1136 'Interactions Arbres/Microorganismes', Champenoux, France; e-mail: claire.fourrey@univ-lorraine.fr

Biology of the Fungal Cell, 3rd Edition
The Mycota VIII
D. Hoffmeister, M. Gressler (Eds.)

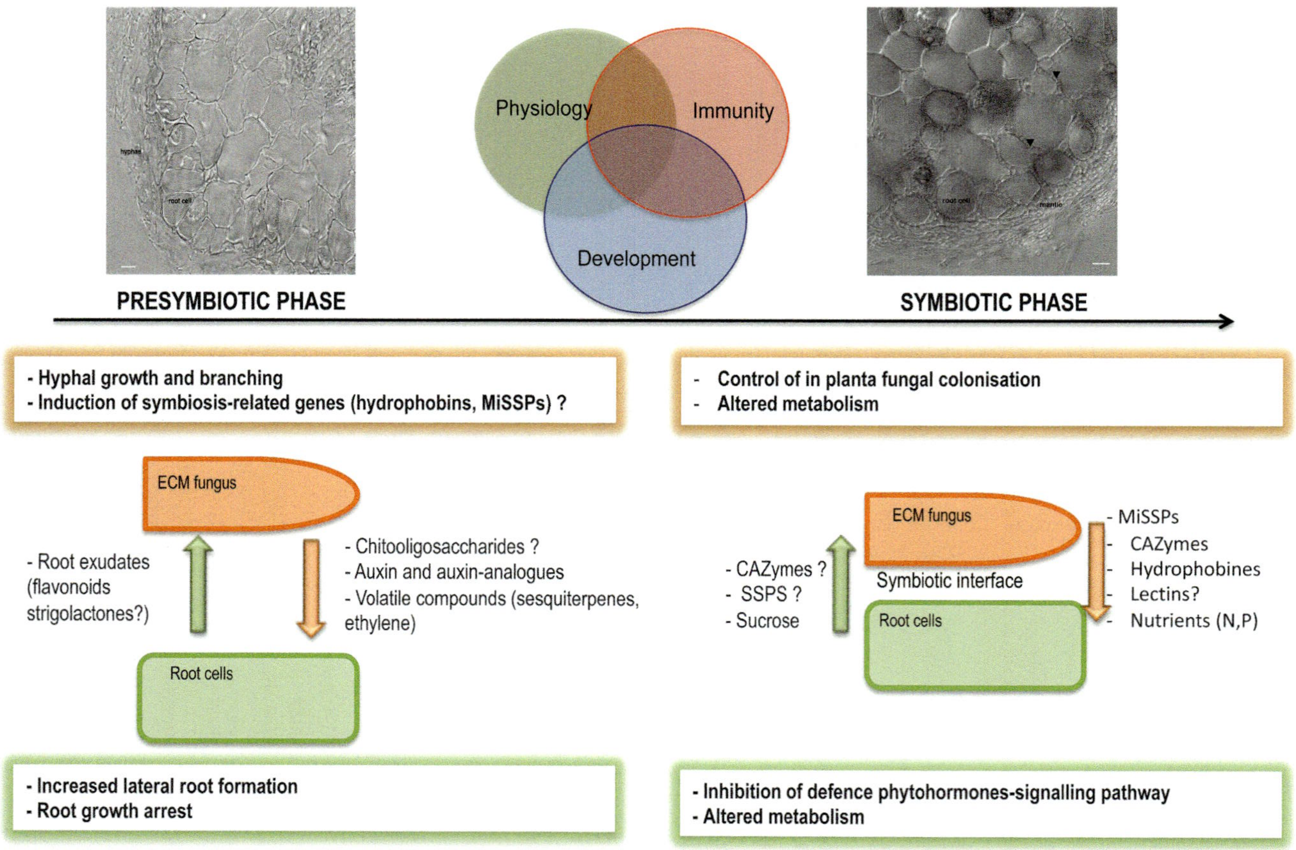

Fig. 1 Molecular dialogue at the ectomycorrhiza symbiotic interface. Several different processes occur during the ontogenesis of ectomycorrhizal symbiosis. Presymbiotic phase is driven by root exudates, volatiles (VOC) and fungal indole acid acetic (IAA). IAA leads to root growth arrest and stimulates fungal hyphae branching. After mantle formation, fungal hyphae penetrate within the apoplast. Plant cell wall-degrading enzymes are secreted into the apoplastic space in order to reshape plant and fungal cell walls and create the symbiotic interface. At the same time, fungus releases secreted proteins likely involved in either symbiotic interface or in reprogramming host physiology, immunity in order to promote symbiosis. Finally, a bidirectional exchange of nutrients is set between the plant and fungus within the Hartig net, leading to a long-term mutualistic symbiosis. The development of a functional ectomycorrhiza thus requires a strict control of fungal and root cells, through secreted signals and their responses. Understanding the biology of ectomycorrhiza is at the overlap of plant and fungal development, immunity and physiology. Arrowheads indicate the symbiotic interface, namely, the Hartig net. Scale bars indicate 20 μm

carbon sequestration through photosynthesis (Sedjo and Sohngen 2012) but also water quality, wood production and protection against soil erosion. Thus, forest ecosystems play a major role in the global climate change context (Hanewinkel et al. 2013), stressing the urgent need to consider land management and tree diversity on a long-term range (Foley et al. 2005). Mycorrhizal fungi play a critical role in forest ecosystems due to their ability to prospect large area of soils with their mycelia, forming a mycorrhizal network that may interconnect roots of trees together (Simard et al. 2012; Pickles and Simard 2016) as well as their role for carbon sequestration within the soils (Clemmensen et al. 2015). Notably, about 6000 of tree species (belonging to Pinaceae, Fagaceae, Betulaceae, Nothofagaceae, Myrtaceae and Dipterocarpaceae) are able to form a mycorrhizal symbiosis with 50,000 ectomycorrhizal (ECM) fungal species (van der Heijden et al. 2015). These ECM fungi participate in turnover of soil nutrients (C, N, P, K, Ca, Fe) through their extraradical mycelium (Marschner and Dell 1994; Ekblad et al. 2013). The ECM symbiosis appeared several times independently in the evolution about 150 million years ago (Hibbett and Matheny 2009; Ryberg and Matheny 2012; Kohler et al. 2015; van der Heijden et al. 2015) and differs significantly from the other mycorrhizal fungi in terms of symbiotic structures, their ecology and their phylogenetic position in the fungal tree of life. Most of the ectomycorrhizal fungi belong to *Basidiomycota*, although some fungal species such as the black truffle of Perigord (*Tuber melanosporum*) belongs to *Ascomycota*. Realizing this prevalence in such ecosystems, it is then critical to decipher and analyse the molecular processes driving the symbiosis between these soilborne fungi and their host plants in order to gain an accurate view on the molecular mechanisms occurring in forest ecosystem. For many years, studies on mycorrhizal mutualistic interactions have been limited by the lack of data on mycorrhizal fungal genomes. Availability of first mycorrhizal genomes and rising of next-generation sequencing (NGS) technology gave us access to hundreds of fungal genomes, among them 50 ECM fungal genomes (Martin et al. 2008, 2010; Martin 2011; Floudas et al. 2012; Kuo et al. 2014; Köhler et al. 2015; Marmeisse et al. 2013). In addition, NGS allows researchers to move on by looking at transcriptome and predicted secretome of those beneficial symbionts. The purpose of this chapter is to give an overview on the ontogenesis of ectomycorrhizal symbiosis and to highlight how comparative genomics and molecular biology lead to reconsider mutualistic interactions as an evolutionary habit sharing traits with saprotrophic and plant-pathogenic fungal species. We will focus in this chapter on (1) the understanding of fungal and plant signalling molecules guiding mycorrhizal interactions and (2) the identification of the underlying receptors and signalling pathways, with a particular emphasis on the control of hormone signalling pathways (Fig. 1). We will finally address the role of secreted proteins in the establishment and/or functioning of ECM symbiosis.

II. The Presymbiotic Phase

A. Plant Rhizospheric Compounds Are Preparing ECM Fungi for Infection

Perception and recognition of signal molecules produced by both trees and ECM fungi is a prerequisite for the establishment of compatible ECM symbiosis. This presymbiotic phase allows the ECM fungus to sense and recognize a putative host partner. Search for host-derived signal molecules able to stimulate spore germination, ECM hyphae branching and chemotropism towards host root led to the identification of only a limited number of compounds.

Fries et al. (1987) first demonstrates that root exudates and extracts from *Pinus sylvestris* (in particular abietic acid) stimulate germination of *Suillus* spp. basidiospores. During the initial stages of in vitro colonization of *Larix decidua* roots by *Suillus tridentinus*, flavonols such as catechin and epicatechin reached more than 50 mM and two minor metabolites, 4-hydroxybenzoate glucoside and quercetin rhamnoside, exhibited a rapid accumulation in the mycorrhiza root tip, suggesting a putative role for pre-infection steps (Weiss et al. 1997).

Plant root exudates and root volatiles are inducing expression of two hydrophobin genes of the ectomycorrhizal fungus *Tricholoma vaccinum* at the presymbiotic stage, followed by the induction of a third hydrophobin gene during mantle formation (Sammer et al. 2016).

The presence of rutin in exudates of *Eucalyptus globulus* spp. *bicostata* stimulates radial growth and biomass of different *Pisolithus* sp. strains (Lagrange et al. 2001).

However, only *Pisolithus* strains collected under rutin-producing *Eucalyptus* are receptive to rutin, suggesting, first, a putative role of flavonoids in specificity of host-fungal interaction and second, that ECM fungi are able to adapt to the chemical signature of the rhizosphere they are growing in.

Strigolactones (SLs) are carotenoid-derived compound and a new group of plant hormones that suppress lateral shoot branching and are crucial to root development. In addition, their role as signalling molecules in the rhizosphere is well documented.

For example, SLs stimulate both spore germination and hyphal growth and branching of *Rhizophagus irregularis*. A recent study suggests SL contribute to arbuscular-mycorrhiza (AM) fungal colonization via the induction of putative-secreted protein SIS1 (Tsuzuki et al. 2016).

Those data suggest a role of SL in both presymbiotic and symbiotic stages for AM symbiosis development. Analyses of available plant genomes and transcriptomes revealed that all analysed plants contain genes required for strigolactone production (Delaux et al. 2012; reviewed in Garcia et al. 2015). However, so far, no effect of strigolactone analogue GR24, on either fungal growth or hyphal branching, was observed in a wide range of soilborne fungi, including ECM fungi (Steinkellner et al. 2007).

AM fungi and *Rhizobia*-nodulating bacteria produce upon perception of plant flavonoids or strigolactones, respectively, Myc- and Nod-factors, sulphated or not lipochitooligosaccharides (Maillet et al. 2011; Denarié and Cullimore 1993). For now, such Myc-factors have not been identified in ECM fungi. We can wonder whether these Myc-factors exist among ECM fungi. Identification of such compounds would be a major achievement in the understanding of host-ECM fungi recognition and host specificity.

To conclude, whereas plant-derived signalling molecules, such as flavonoids and strigolactones, are important molecules for Rhizobium-legumes interactions and AMF-plant interactions, the plant-derived signalling molecules required for ECM establishment (pre-contact phase) are still largely unknown and likely diverse. One can then wonder whether the chemical dialogues involved are host species specific or not.

B. ECM Fungal-Derived Signalling Molecules Alter Host Root Architecture and Development

Root development and architecture are profoundly modified in the vicinity of fungal hyphae and before direct contact, to form ECM root tips. Indeed, increased lateral root development and root growth arrest (Dexheimer and Pargney 1991; Ditengou et al. 2000; Felten et al. 2009; Spivallo et al. 2009; Rincon et al. 2003; Vayssières et al. 2015) are observed, suggesting that root cells respond to fungal signalling molecules.

1. Fungal Auxin as an Early Diffusible Signalling Molecule

Several authors suggested that fungal auxin has an important role as a signalling molecule to initiate ECM development (Gay et al. 1994; Rincon et al. 2001; Felten et al. 2010; Spivallo et al. 2009). Auxin and auxin-analogues released from several ECM fungi were found (Niemi et al. 2002; Krause et al. 2015; Vayssières et al. 2015).

Pisolithus microcarpus is able to produce the indolic compound called hypaphorine, a tryptophan betaine (Béguiristain and Lapeyrie 1997). Krause et al. (2015) identified the exporter of IAA from *Tricholoma vaccinum* as the multidrug and toxic extrusion (MATE) transporter Mte1, supporting the efflux of IAA from fungal cells. Several studies show an increased level of auxin in mature ECM of *Populus trichocarpa-Laccaria bicolor* (Vayssières et al. 2015), *Hebeloma-Pinus* and *Piloderma-Quercus* (Hermann et al. 2004),

highlighting that ECM root tips are auxin-rich environment. As fungal auxins stimulate fungal cell elongation and hyphal branching from ECM fungi and not from non-mycorrhizal fungi (Debaud and Gay 1987; Krause et al. 2015), fungal IAA is likely also required for the observed hyphae hyperbranching phenotype at both early and later stages of fungal colonization. At early stage, hyperbranching is likely enhancing the probability of fungal hyphae to encounter roots and at later stage a requisite for hyphal network formation (known as the Hartig net) (Krause et al. 2015). Hypaphorine from *Pisolithus microcarpus* is able to trigger the expression of auxin-signalling genes and regulate auxin level in *Eucalyptus globulus Labill.* ssp. *bicostata* roots and could regulate auxin transport (Béguiristain and Lapeyrie 1997; Ditengou et al. 2000; Nehls et al. 1998). In the *Laccaria bicolor-Populus* interaction, lateral root stimulation at early stages of ECM establishment is achieved by modulating auxin transporter, in particular by enhancing the basipetal auxin transport (Felten et al. 2009). This is achieved likely after the concentration of auxin in root apices increased (Felten et al. 2009; Vayssières et al. 2015).

To conclude, the current working model is as follows: fungal auxins are synthesized in hyphae in the vicinity of roots, secreted and transferred to root cells where it triggers host auxin signalling as well as activates auxin production, leading into **root growth arrest** and increased **lateral root development** (Vayssières et al. 2015).

2. ECM Fungi Produce Volatile Organic Compounds as Signalling Molecules to Stimulate Lateral Root Development

Interestingly, stimulation of lateral roots (LR) is also observed in the non-host plant *Arabidopsis thaliana* in indirect contact with either *L. bicolor* or *Tuber melanosporum* (Spivallo et al. 2009; Felten et al. 2009, 2010). This suggests fungal production of volatiles or diffusible compounds used as signal molecules by the plant.

Interestingly, mutant *ein2* of *A. thaliana* is less sensitive to LR stimulation, suggesting that ethylene is likely involved. In addition, some but not all ECM fungi are able to produce ethylene (Graham and Linderman 1980; Rupp and Mudge 1985; Splivallo et al. 2009), whereas others are likely able to produce jasmonates (Regvar et al. 1997) or cytokinins (Kraigher et al. 1991; Kovač and Žel 1995).

Ethylene and jasmonic acid when applied exogenously induce LR development (Swarup et al. 2007; Ivanchenko et al. 2008; Sun et al. 2009), whereas these two hormones act as negative regulators of ECM symbiosis establishment (Plett et al. 2014b). A recent study demonstrated that ectomycorrhizal symbiont *L. bicolor* diffuse at very low concentration of thujopsene, a volatile organic compound belonging to the sesquiterpene family, prior any physical contact with the plant (either *P. trichocarpa* or the non-mycorrhizal plant *A. thaliana*) in order to trigger increased lateral root through a ROS-dependent mechanism (Ditengou et al. 2015). *Laccaria* sesquiterpenes likely do not target auxin signalling as they do not modify the auxin reporter DR5::GFP signal intensity and spatial distribution (Ditengou et al. 2015). This study highlights that ECM fungi produce volatiles as many other microbes able to interact with plants, suggesting a common strategy despite the different chemical nature of volatiles (Junker and Tholl 2013; Bailly et al. 2014; Bitas et al. 2013; Schmidt et al. 2016).

In conclusion, several ECM fungi, by releasing **volatiles** (e.g. sesquiterpenes) and/or **hormones** (IAA, ethylene, cytokinin), stimulate lateral roots development leading to more root surface area to colonize. Several molecules (e.g. fungal IAA) are acting through the manipulation of auxin gradient and/or transport, whereas other such as sesquiterpenes act independently of auxin. In addition, in some cases IAA also stimulates branching of fungal hyphae to enhance both the contact between roots and hyphae and later the Hartig net development. However, nothing yet proves this strategy is broadly used by ECM fungi in the early steps of interaction. The future identification of more fungal-derived signal molecules would elucidate whether a common molecule is required for ECM symbiosis establishment.

III. Control of Plant Hormone Signalling Pathways in Later Stages of ECM Development

After hyphae and roots enter in contact, fungal hyphae completely surround small lateral roots and will grow intensively within the apoplast of cortical cells to form the Hartig net. At this stage, phytohormone networks are likely the modulators of root development, nutrient homeostasis (sustain bidirectional nutrient fluxes) and immunity. A recent review (Pozo et al. 2015) highlights that nearly all studied phytohormones so far play a role in AM establishment or functioning. Several studies also showed that ethylene (ET), jasmonic acid (JA) and salicylic acid (SA) are produced and sometimes accumulated during the interaction between mutualistic fungi and their host plants (Hause et al. 2007; Splivallo et al. 2009; Lahmann et al. 2015; Khatabi and Schäfer 2012). Targeting hormonal networks of plants (biosynthesis and signalling) either by the production of either secreted proteins or fungus-derived phytohormone mimics is common to both pathogenic and mutualistic fungi (for a recent review, Shen et al. 2018).

A. Auxin

The phytohormone auxin (indole-3-acetic acid [IAA]), known for its central and pleiotropic roles in plant growth and metabolism (Zhao 2010; Teale et al. 2006), is considered to be one of the main triggers regulating all of the different steps of LR formation (Ditengou et al. 2008; Laskowski et al. 2008; Nibau et al. 2008). In the previous section, we tackled the role of fungal IAA at early stages of ECM development. In this paragraph, we will discuss the role of auxin in later steps of ECM development. We stated in a previous section that fungal IAA perception likely regulates auxin transport genes in roots. In addition to modulation of auxin transport, ECM development is also altering auxin metabolism (in particular amino-acid conjugation is enhanced, Vayssières et al. 2015) and auxin signalling. For example, in the early stages of interaction between either *L. bicolor* or *H. cylindrosporum* and pine, auxin-signalling genes were induced (Charvet-Candela et al. 2002; Heller et al. 2012). Similarly, the interaction between *Pisolithus tinctorius* and oak roots led to an induction of auxin biosynthesis and auxin-signalling-related genes (Sebastiana et al. 2014). Interestingly, in the late stages of the interaction between *Piloderma* and oak, auxin transporters and auxin-responsive transcription factors are repressed, suggesting that auxin gradients are only important in the early stages of ECM mycorrhizal root formation (Tarkka et al. 2013).

Hebeloma cylindrosporum transgenic fungus over-producing IAA is able to better colonize roots, with the number of colonized roots is increasing by 3–5 times (Gay et al. 1994). Microscopic analysis of ECM formed by the over-producer of IAA of *H. cylindrosporum* displayed a denser Hartig-net, suggesting a role of fungal IAA at early and late stages of ECM development.

Gay et al. (1994) proposed that fungal IAA might play a role in cell wall loosening at this later stage, whereas Krause et al. (2015) proposed that IAA is stimulating hyphae hyper-branching, necessary to form the pseudo-parenchymatous tissues, the Hartig net.

From their studies and others, Vayssières et al. (2015) proposed a working model: (1) auxin-rich environment in root apices would be achieved partly through fungal IAA uptake; (2) auxin-signalling pathway would be triggered leading to more lateral roots; (3) concentration of IAA conjugate to amino acids would increase; and (4) part of the auxin responses would be repressed in mature *L. bicolor-Populus* ECM root tips suggesting either a plant autoregulation of auxin signalling or the action of a fungal protein able to interact with auxin signalling (Vayssières et al. 2015).

B. Ethylene

Using transgenic *Populus tremula x tremuloides*, Plett et al. (2014b) demonstrated that neither ACC overproduction (*Populus* overexpressing ACOI1) nor ET insensitivity (*Populus etr1-1* mutants) affects the initial stages of ECM development, namely, the mantle formation. However, both *Poplar* plantlets overexpressing ACO1 and treatment with exogenous ACC, the precursor of ET, inhibit fungal colonization of the apoplast of rhizodermal cells, i.e. the formation of Hartig net (Plett et al. 2014b).

Compromised ethylene-signalling pathway in *Arabidopsis thaliana* also led to a greater fungal colonization by the root endophyte *Piriformospora indica* (Camehl et al. 2010). More precisely, Ethylene Response Factor (ERF1), an ethylene-signalling relay protein, is likely the key regulator in both *Populus-L. bicolor* and *A. thaliana-P. indica* interactions, as up-regulation of either AtERF1 or PtaERF1 resulted in a decrease fungal colonisation of rhizodermal cells (Camehl et al. 2010; Plett et al. 2014b). However, another study showed an opposite result in which *A. thaliana* impaired in ethylene signalling displayed a reduced *P. indica* root colonization, whereas *Arabidopsis* mutants exhibiting constitutive ethylene signalling, ethylene synthesis or ethylene-related defence displayed a better fungal colonization (Khatabi and Schäfer 2012).

This suggests that the role of ET may also depend on growth and infection conditions. We can hypothesize from these studies that ET inhibits Hartig net formation late stages of fungal colonization but not primary stages of ECM development mantle formation. Experiments conducted on other host-ECM fungi associations are required to assess whether the role of ET in controlling symbiosis development is conserved.

Application of exogenous ACC or overexpression of PtACO1 resulted in the repression of genes associated with cell wall biosynthesis (expansin, pectinesterase, cellulose synthase), suggesting a role of ET in controlling plant cell wall composition during fungal colonization (Plett et al. 2014b).

Ethylene-responsive genes are induced at very late stages of colonization as showed by several transcriptomic studies (Plett et al. 2014b; Tarkka et al. 2013), suggesting that ET-signalling is triggered only at late stages of ECM development. However, Spivallo et al. (2009) showed that *A. thaliana ein2* mutant, insensitive to ET, form less lateral roots than the wild-type under indirect contact with *Tuber melanosporum*. As they showed that *Tuber melanosporum* synthesize ethylene from l-methionine probably through the α-keto-γ-(methylthio)butyric acid pathway and concluded that ET signalling triggered by fungal ethylene is essential to cooperate with auxin at early stages of ECM establishment.

From these studies, it can be suggested that (1) fungal ethylene is required at earliest stages of ECM development and acts cooperatively with plant auxins to stimulate LR formation and (2) plant ethylene, while not required at the early stages of the symbiosis, is needed to limit fungal growth within root tissues.

C. Jasmonic Acid

Treatment of *Populus trichocarpa* roots by jasmonic acid (JA) induces a drastic decrease of *L. bicolor* hyphae penetration, whereas the formation of the mantle is not influenced (Plett et al. 2014b). However, JA is produced when *L. bicolor* starts to penetrate the apoplastic space between rhizodermal cells (Veneault-Fourrey and Schellenberger, unpublished results). However, in mature ECM root tips, JA content is decreased when compared to non-ECM roots of grey poplar (*Populus tremula* x *Populus alba*) (Luo et al. 2009).

On the other hand, colonisation of *A. thaliana* roots by the fungal endophyte *P. indica* also increases JA level in roots. JA treatment of *Populus* roots colonized by *L. bicolor* triggers the expression of stress-related genes (e.g. thaumatin-related genes, receptor-like kinases, chitinases), expansin encoding genes and represses the expression of cell wall active enzymes (e.g. pectin esterases, xyloglucan transferases, pectinases, expansin) (Plett et al. 2014b). This suggests an increased resistance to fungal colonization and a decrease of cell elongation (Cho and Cosgrove 2000; Hall and Cannon 2002; Humphrey et al. 2007). Transcriptomic data showed an extensive cross-talk between ET and JA transcriptomic changes in ECM root tips of *Populus trichocarpa* (Plett et al. 2014b).

Therefore, it appears that JA and ET are required to set boundaries to fungal colonization through a cooperative regulation of **cell wall remodelling** and triggering of **plant defences** (Plett et al. 2014b). It is thus likely that ECM fungi manipulate the local plant defence responses to create a localized weakening of the plant immune system. Some mutualistic fungi as pathogenic fungi use secreted proteins to modulate the impact of these hormones and were consequently named **effectors** (Plett et al. 2014b; Kloppholz et al. 2011; Shen et al. 2018).

IV. Secreted Proteins, Central Players in Controlling Symbiosis Development?

First genome-wide analysis carried out on *L. bicolor-Populus trichocarpa* ECM root tips revealed a contrasting transcriptomic landscape with the free-living mycelium. Among the 20 most up-regulated genes during the symbiosis, 5 of them code for secreted proteins SSPs (Martin et al. 2008). Due to their strong up-regulation during symbiosis and their size, they were so-called mycorrhiza-induced small secreted proteins (MiSSPs). At this time, it has been hypothesized that these MiSSPs could be symbiotic effectors, with the definition of effector as "microbial secreted molecules that alter host-cell processes or structure generally promoting the microbe lifestyle" (Win et al. 2012). *L. bicolor* MiSSP7, whose encoding gene is one of the most up-regulated during symbiosis, is secreted and enters host cells where it localizes into the nucleus (Plett et al. 2011).

Inside the nucleus, MiSSP7 interacts with the poplar PtJAZ6 protein (Plett et al. 2014a). JAZ proteins are negative regulator of jasmonic acid signalling pathway. Once jasmonic acid is produced, JAZ proteins are ubiquitinated and degraded in a proteasome-dependent manner (Pauwels and Goossens 2011; Wager and Browse 2012). MiSSP7 stabilizes PtJAZ6 avoiding its degradation in the presence of JA. Transcription of target genes is thus repressed. Preliminary results showed that genes involved in plant cell wall remodelling and plant defence responses might be target genes of this protein complex (Plett et al. 2014a).

These studies together with previous ones on plant-pathogenic microbes (Jiang et al. 2013; Gimenez-Ibanez et al. 2014) highlight that JAZ proteins are cellular hubs targeted by plant-interacting microbes to facilitate in planta colonization. However, whereas bacterial effectors stimulate JA signalling pathway, mutualistic fungi are decreasing JA signalling pathway.

Availability of *Rhizophagus irregularis* (formerly *Glomus intraradices*) genome sheds the light on several secreted proteins whose encoding genes were up-regulated during the AM symbiosis with *Medicago truncatula*.

Among them, SP7 has been shown to interact with MtERF19, a transcription factor from *Medicago truncatula* involved in plant immune responses (Kloppholtz et al. 2011). By expressing SP7-encoding genes in the rice blast fungus, *Magnaporthe oryzae*, it extends the biotrophic phase of this pathogenic fungus, demonstrating the ability of a secreted protein from a mutualistic fungus to impact life cycle of fungi from other lifestyles. The root endophytic fungus *Piriformospora indica* is also secreting PIIN_08944 needed for *Arabidopsis thaliana* root colonization with potential interference with burst of reactive oxygen species and salicylic acid-mediated immune responses (Akum et al. 2015). The *P. indica* FGB1 protein is a secreted fungal-specific β-glucan-binding lectin with dual function.

FGB1 affects fungal cell wall composition, and its over-expression increases resistance to cell wall-stress-inducing compounds. FGB1 also suppresses β-glucan-triggered immunity (Warwa et al. 2016).

Interestingly, the *L. bicolor* GH5-CBM1 protein is a symbiosis-induced endoglucanase displaying activity towards cellulose, mannans and galactomannans without activity against fungal cell walls.

L. bicolor GH5-CBM1 RNAi mutants are impaired in their symbiosis ability and the protein localizes at the periphery of hyphal cells from both the mantle and the Hartig net.

These data suggest a role of fungal secreted enzyme involved in plant cell wall remodelling in the establishment of ectomycorrhiza (Zhang et al. 2018). *L. bicolor missp8* RNAi knockdown mutants display disorganized fungal hyphae and are consequently impaired in fungal mantle and Hartig net development (Pellegrin et al. 2017). In the ericoid mycorrhizal fungus *Oidiodendron maius*, the most highly induced secreted proteins in symbiosis OmSSP1 is required for mycorrhization and is a hydrophobin-like fungal protein (Casarrubia et al. 2018).

These data reinforce the supposed structural role of fungal proteins in mutualistic interactions such as *P. tinctorius* SRAP32 (Laurent et al. 1999) and hydrophobins (Tagu et al. 1998).

Those recent functional studies point out a role of **secreted proteins** in (1) the maintenance of both plant and fungal cell wall integrity to sustain the formation of symbiotic interface and (2) the sensing of cell wall integrity likely to avoid triggering of immune responses.

When comparing the expression of MiSSP genes of *L. bicolor* in mutualistic interaction with *P. trichocarpa* (Angiosperm) and with *Pseudotsuga menziesii* (Gymnosperm), a core regulon and a variable regulon of MiSSPs were identified (Plett et al. 2014b). The variable regulon may explain host specificity. In parallel, proteomic analysis of *L. bicolor* free-living mycelium revealed *L. bicolor* secretes both cell wall degrading enzymes and proteases as well as nine secreted proteins (Vincent et al. 2012).

Computational analysis of *Hebeloma cylindrosporum* genome combined with proteomic-based study of the free-living mycelium exoproteome leads to the identification of 869 secreted proteins (Doré et al. 2015). A large part (43%) of the proteins of the secretome lacks functional annotation (Doré et al. 2015). Interestingly, even though some SSPs are up-regulated during the symbiosis, 28% of SSPs were expressed in the free-living mycelia (Doré et al. 2015). These latter SSPs are mostly well conserved across fungal kingdom (Doré et al. 2015).

Such SSPs repressed in ECM and/or expressed in FLM and conserved across fungal kingdom could have a role in fungal biology likely related to the saprotrophic/decomposer stage independent from symbiosis (Doré et al. 2015). Doré et al. also compared the expression of SSPs in ECM root tips formed in vitro in synthetic medium with ECM root tips formed in greenhouse on organic matter. Several SSPs expressed in in vitro ECM root tips were poorly expressed in greenhouse ECM root tips and vice versa. These two experiments indicate that SSP-encoding genes may respond not only to host but also to environmental signals. Doré et al. proposed that SSPs are integrator of both biotic and abiotic soil environment. This will be in accordance with the emerging idea that SSP of plant pathogenic fungi may be used also to perform dialogue and competition with other soil microbes (Rovenich et al. 2014). Moreover, in silico comparative analysis of fungal secretomes, including 28 saprotrophic and 12 ECM fungal species, showed that saprotrophic fungi possess hundreds of small proteins predicted as secreted, demonstrating the presence of SSPs is not restricted to biotrophic microorganisms (Pellegrin et al. 2015). Clustering analysis also identified clusters of SSPs shared between saprotrophic and ECM fungi, raising the hypothesis of a common repertoire of SSPs between saprotrophic and ECM fungi. However, it is difficult to assess their importance in fungal lifestyle due to the lack of transcriptomic and functional data.

Altogether, these results suggest that mutualistic fungi secrete small proteins to manipulate their hosts in particular by targeting defence phytohormone signalling pathways or interfering with plant immunity or modifying the structure of either fungal or plant cell walls. It is important to note that the expression of MiSSPs has also been found in all mycorrhizal interactions studied so far (Kohler et al. 2015) and also in situ in *Piloderma croceum-Pinus taeda* ectomycorrhizal root tips (Liao et al. 2014). However, available literature is still very poor, and functional analyses of new secreted proteins are required to clarify and detail how mycorrhizal fungi coordinate their activity with their host plant (Martin et al. 2016). In addition, understanding how these secreted proteins have evolved from their saprotrophic ancestors is one of the next challenges.

V. Perspectives and Future Research

Availability of –omics data provided an effective stepping-stone for functional analysis.

Combination of transcriptional and functional analysis sheds the light on MiSSPs for both arbuscular and ectomycorrhizal fungi. This discovery has large consequences, as pathogenic interactions (and N-fixing rhizobacteria) were the only one known to employ secreted proteins to manipulate their host and promote the interaction.

It is now a challenge to decipher the functions of these secreted proteins considering their lack of similarities with proteins with known function. However, this step forward is necessary to understand the development and functioning of ectomycorrhizal symbiotic interaction and gain a better view on the molecular dialogue in plant-microbe interactions.

Furthermore, the presence of secreted proteins expressed during free-living stages of the ectomycorrhizal symbionts *H. cylindrosporum* and *L. bicolor* indicates that ECM fungi use secreted proteins to fulfill both symbiotic and non-symbiotic functions (Doré et al. 2015; Plett et al. 2015). Therefore, it will be of great interest to study the role of secreted proteins in saprotrophic fungal species and in the free-living mycelium of ECM fungal species to test whether they have similar function or whether neofunctionalization occurred. Although our understanding of ECM symbiosis has greatly increased in the last 10 years, there are still many questions to address to understand the symbiotic signalling pathways:

- To what extend environment and host affect evolution of ECM fungal genomes?
- Are secreted proteins a widespread strategy within mutualistic interactions and how do they impact the host development and metabolism?
- What are the master regulators of fungal MiSSPs-encoding genes?
- Most importantly, what are the host proteins and functions targeted by the MiSSPs? Are there "hubs" targeted by secreted proteins from different ECM fungal species?
- Is the control of hormonal pathways involved in plant immunity by fungal MiSSPs local or systemic? How could this impact other kind of on plant-microbe interactions?
- How does the host secretome participate to the molecular dialogue established with the fungus?

All these questions are challenging as each ECM fungus seems to possess its own symbiotic toolbox, but combination of –omics technologies and functional analysis will help dissecting of this belowground symbiotic relationships.

References

Akum FN, Steinbrenner J, Biedenkopf D, Imani J, Kogel K-H (2015) The *Piriformospora indica* effector PIIN_08944 promotes the mutualistic Sebacinalean symbiosis. Front Plant Sci 6:906

Bailly A, Groenhagen U, Schulz S, Geisler M, Eberl L, Weisskopf L (2014) The inter-kingdom volatile signal indole promotes root development by interfering with auxin signaling. Plant J 80:758–771

Béguiristain T, Lapeyrie F (1997) Host plant stimulates hypaphorine accumulation in Pisolithus tinctorius hyphae during ectomycorrhizal infection while excreted fungal hypaphorine controls root hair development. New Phytol 136:525–532

Bitas V, Kim H-S, Bennett JW, Kang S (2013) Sniffing on microbes: diverse roles of microbial volatile organic compounds in plant health. Mol Plant Microbe Interact 26:835–843

Camehl I, Sherameti I, Venus Y, Bethke G, Varma A, Lee J, Oelmüller R (2010) Ethylene signalling and ethylene-targeted transcription factors are required to balance beneficial and non beneficial traits in the symbiosis between the endophytic fungus *Piriformospora indica* and *Arabidopsis thaliana*. New Phytol 185:1062–1073

Casarrubia S, Daghino S, Kohler A, Morin E, Khouja H-R, Daguerre Y, Veneault-Fourrey C, Martin FM, Perotto S, Martino E (2018) The hydrophobin-like OmSSP1 may be an effector in the ericoid mycorrhizal symbiosis. bioRxiv. https://doi.org/10.1101/227959

Charvet-Candela V, Hitchin S, Reddy MS, Cournoyer B, Marmeisse R, Gay G (2002) Characterization of a *Pinus pinaster* cDNA encoding an auxin up-regulated putative peroxidase in roots. Tree Physiol 22:231–238

Cho H-T, Cosgrove DJ (2000) Altered expression of expansin modulates leaf growth and pedicel abscission in *Arabidopsis thaliana*. Proc Natl Acad Sci 97:9783–9788

Clemmensen KE, Finlay RD, Dahlberg A, Stenlid J, Wardle DA, Lindahl BD (2015) Carbon sequestration is related to mycorrhizal fungal community shifts during long-term succession in boreal forests. New Phytol 205(4):1525–1536

Debaud JC, Gay G (1987) *In vitro* fruiting under controlled conditions of the ectomycorrhizal fungus *Hebeloma cylindrosporum* associated with *Pinus pinaster*. New Phytol 105:429–435

Delaux PM, Xie X, Timme RE, Puech-Pages V, Dunand C, Lecompte E, Delwiche CF, Yoneyama K, Bécard G, Séjalon-Delmas N (2012) Origin of strigolactones in the green lineage. New Phytol 195:857–871

Dénarié J, Cullimore J (1993) Lipo-oligosaccharide nodulation factors: a new class of signaling molecules mediating recognition and morphogenesis. Cell 74:951–954

Dexheimer J, Pargney JC (1991) Comparative anatomy of the host-fungus interface in mycorrhizas. Experientia 47:312–321

Ditengou F, Béguiristain T, Lapeyrie F (2000) Root hair elongation is inhibited by hypaphorine, the indole alkaloid from the ectomycorrhizal fungus *Pisolithus tinctorius*, and restored by indole-3-acetic acid. Planta 211:722–728

Ditengou FA, Teale WD, Kochersperger P, Flittner KA, Kneuper I, van der Graaff E, Nziengui H, Pinosa F, Li X, Nitschke R, Laux T, Palme K (2008) Mechanical induction of lateral root initiation in *Arabidopsis thaliana*. Proc Natl Acad Sci 105:18818–18823

Ditengou FA, Müller A, Rosenkranz M, Felten J, Lasok H, Van Doorn MM, Legué V, Palme K, Schnitzler JP, Polle A (2015) Volatile signalling by sesquiterpenes from ectomycorrhizal fungi reprogrammes root architecture. Nat Commun 6. https://doi.org/10.1038/ncomms7279

Doré J, Perraud M, Dieryckx C, Kohler A, Morin E, Henrissat B, Lindquist E, Zimmermann SD, Girard V, Kuo A, Grigoriev IV, Martin F, Marmeisse R, Gay G (2015) Comparative genomics, proteomics and transcriptomics give new insight into the exoproteome of the basidiomycete *Hebeloma cylindrosporum* and its involvement in ectomycorrhizal symbiosis. New Phytol 208:1169–1187

Ekblad A, Wallander H, Godbold DL, Cruz C, Johnson D, Baldrian P, Björk R, Epron D, Kieliszewska-Rokicka B, Kjøller R, Kraigher H, Matzner E, Neumann J, Plassard C (2013) The production and turnover of extramatrical mycelium of ectomycorrhizal fungi in forest soils: role in carbon cycling. Plant Soil 366:1–27

Felten J, Kohler A, Morin E, Bhalerao RP, Palme K, Martin F, Ditengou FA, Legue V (2009) The ectomycorrhizal fungus *Laccaria bicolor* stimulates lateral root formation in poplar and Arabidopsis through auxin transport and signaling. Plant Physiol 151:1991–2005

Felten J, Legué V, Ditengou FA (2010) Lateral root stimulation in the early interaction between *Arabidopsis thaliana* and the ectomycorrhizal fungus *Laccaria bicolor*: is fungal auxin the trigger? Plant Signal Behav 5:864–867

Floudas D, Binder M, Riley R, Barry K, Blanchette RA, Henrissat B, Martínez AT, Otillar R, Spatafora JW, Yadav JS, Aerts A, Benoit I, Boyd A, Carlson A, Copeland A, Coutinho PM, De Vries RP, Ferreira P, Findley K, Foster B, Gaskell J, Glotzer D, Górecki P, Heitman J, Hesse C, Hori C, Igarashi K, Jurgens JA, Kallen N, Kersten P, Kohler A, Kües U, Kumar TKA, Kuo A, LaButti K, Larrondo LF, Lindquist E, Ling A, Lombard V, Lucas S, Lundell T, Martin R, McLaughlin DJ, Morgenstern I, Morin E, Murat C, Nagy LG, Nolan M, Ohm RA, Patyshakuliyeva A, Rokas A, Ruiz-Dueñas FJ, Sabat G, Salamov A, Samejima M, Schmutz J, Slot JC, John FS, Stenlid J, Sun H, Sun S, Syed K, Tsang A, Wiebenga A, Young D, Pisabarro A, Eastwood DC, Martin F, Cullen D, Grigoriev IV, Hibbett DS (2012) The paleozoic origin of enzymatic lignin decomposition reconstructed from 31 fungal genomes. Science 80-336:1715–1719

Foley JA, DeFries R, Asner GP, Barford C, Bonan G, Carpenter SR, Chapin FS, Coe MT, Daily GC, Gibbs HK, Helkowski JH, Holloway T, Howard EA, Kucharik CJ, Monfreda C, Patz JA, Prentice IC, Ramankutty N, Snyder PK (2005) Global consequences of land use. Science 80-309:570–574

Fries N, Serck-Hanssen K, Dimberg LH, Theander O (1987) Abietic acid, and activator of basidiospore germination in ectomycorrhizal species of the genus *Suillus* (Boletaceae). Exp Mycol 11:360–363

Garcia K, Delaux PM, Cope KR, Ané JM (2015) Molecular signals required for the establishment and maintenance of ectomycorrhizal symbioses. New Phytol 208:79–87

Gay G, Normand L, Marmeisse R, Sotta B, Debaud JC (1994) Auxin overproducer mutants of *Hebeloma cylindrosporum Romagnesi* have increased mycorrhizal activity. New Phytol 128:645–657

Gimenez-Ibanez S, Boter M, Fernández-Barbero G, Chini A, Rathjen JP, Solano R (2014) The bacterial effector HopX1 targets JAZ transcriptional repressors to activate jasmonate signaling and promote infection in *Arabidopsis*. PLoS Biol 12(2):e1001792

Graham JH, Linderman RG (1980) Ethylene production by ectomycorrhizal fungi, *Fusarium oxysporum* f sp. *pini*, and by aseptically synthesized ectomycorrhizae and *Fusarium*-infected Douglas-fir roots. Can J Microbiol 26:1340–1347

Hall Q, Cannon C (2002) The cell wall hydroxyproline-rich glycoprotein RSH is essential for normal embryo development in *Arabidopsis*. Plant Cell 14:1161–1172

Hanewinkel M, Cullmann D, Schelhaas M-J, Nabuurs G-J, Zimmermann N (2013) Climate change may

cause severe loss in the economic value of European forest land. Nat Clim Chang 3:203–207
Hause B, Mrosk C, Isayenkov S, Strack D (2007) Jasmonates in arbuscular mycorrhizal interactions. Phytochemistry 68:101–110
Heller G, Lundén K, Finlay RD, Asiegbu FO, Elfstrand M (2012) Expression analysis of Clavata1-like and Nodulin21-like genes from *Pinus sylvestris* during ectomycorrhiza formation. Mycorrhiza 22:271–277
Herrmann S, Oelmüller R, Buscot F (2004) Manipulation of the onset of ectomycorrhiza formation by indole-3-acetic acid, activated charcoal or relative humidity in the association between oak microcuttings and *Piloderma croceum*: influence on plant development and photosynthesis. J Plant Physiol 161:509–517
Hibbett DS, Matheny PB (2009) The relative ages of ectomycorrhizal mushrooms and their plant hosts estimated using Bayesian relaxed molecular clock analyses. BMC Biol 7:13. https://doi.org/10.1186/1741-7007-7-13
Humphrey TV, Bonetta DT, Goring DR (2007) Sentinels at the wall: cell wall receptors and sensors. New Phytol 176:7–21
Isbell F, Calcagno V, Hector A, Connolly J, Harpole WS, Reich PB, Scherer-Lorenzen M, Schmid B, Tilman D, Van Ruijven J, Weigelt A, Wilsey BJ, Zavaleta ES, Loreau M (2011) High plant diversity is needed to maintain ecosystem services. Nature 477:199–202
Ivanchenko MG, Muday GK, Dubrovsky JG (2008) Ethylene-auxin interactions regulate lateral root initiation and emergence in *Arabidopsis thaliana*. Plant J 55:335–347
Jiang S, Yao J, Ma KW, Zhou H, Song J, He SY, Ma W (2013) Bacterial effector activates jasmonate signaling by directly targeting JAZ transcriptional repressors. PLoS Pathog 9(10):e1003715
Junker RR, Tholl D (2013) Volatile organic compound mediated interactions at the plant-microbe interface. J Chem Ecol 39:810–825
Khatabi B, Schäfer P (2012) Ethylene in mutualistic symbioses. Plant Signal Behav 7:1634–1638
Kloppholz S, Kuhn H, Requena N (2011) A secreted fungal effector of *Glomus intraradices* promotes symbiotic biotrophy. Curr Biol 21:1204–1209
Kohler A, Kuo A, Nagy LG, Morin E, Barry KW, Buscot F, Canbäck B, Choi C, Cichocki N, Clum A, Colpaert J, Copeland A, Costa MD, Doré J, Floudas D, Gay G, Girlanda M, Henrissat B, Herrmann S, Hess J, Högberg N, Johansson T, Khouja HR, Labutti K, Lahrmann U, Levasseur A, Lindquist EA, Lipzen A, Marmeisse R, Martino E, Murat C, Ngan CY, Nehls U, Plett JM, Pringle A, Ohm RA, Perotto S, Peter M, Riley R, Rineau F, Ruytinx J, Salamov A, Shah F, Sun H, Tarkka M, Tritt A, Veneault-Fourrey C, Zuccaro A, Tunlid A, Grigoriev IV, Hibbett DS, Martin F (2015) Convergent losses of decay mechanisms and rapid turnover of symbiosis genes in mycorrhizal mutualists. Nat Genet 47:410–415
Kovač M, Žel J (1995) The effect of aluminum on cytokinins in the mycelia of *Amanita muscaria*. J Plant Growth Regul 14:117–120
Kraigher H, Grayling A, Wang TL, Hanke DE (1991) Cytokinin production by two ectomycorrhizal fungi in liquid culture. Phytochemistry 30:2249–2254
Krause K, Henke C, Asiimwe T, Ulbricht A, Klemmer S, Schachtschabel D, Boland W, Kothe E (2015) Biosynthesis and secretion of indole-3-acetic acid and its morphological effects on *Tricholoma vaccinum*-spruce ectomycorrhiza. Appl Environ Microbiol 81:7003–7011
Kuo A, Kohler A, Martin FM, Grigoriev IV (2014) Expanding genomics of mycorrhizal symbiosis. Front Microbiol 5:582. https://doi.org/10.3389/fmicb.2014.00582
Lagrange H, Jay-Allgmand C, Lapeyrie F (2001) Rutin, the phenolglycoside from eucalyptus root exudates, stimulates *Pisolithus* hyphal growth at picomolar concentrations. New Phytol 149:349–355
Lahrmann U, Strehmel N, Langen G, Frerigmann H, Leson L, Ding Y, Scheel D, Herklotz S, Hilbert M, Zuccaro A (2015) Mutualistic root endophytism is not associated with the reduction of saprotrophic traits and requires a noncompromised plant innate immunity. New Phytol 207:841–857
Laskowski M, Grieneisen VA, Hofhuis H, Ten Hove CA, Hogeweg P, Marée AFM, Scheres B (2008) Root system architecture from coupling cell shape to auxin transport. PLoS Biol 6:2721–2735
Laurent P, Voiblet C, Tagu D, de Carvalho D, Nehls U, De Bellis R, Balestrini R, Bauw G, Bonfante P, Martin F (1999) A novel class of ectomycorrhiza-regulated cell wall polypeptides in Pisolithus tinctorius. MPMI 12(10):862–871
Liao HL, Chen Y, Bruns TD, Peay KG, Taylor JW, Branco S, Talbot JM, Vilgalys R (2014) Metatranscriptomic analysis of ectomycorrhizal roots reveals genes associated with *Piloderma-Pinus* symbiosis: improved methodologies for assessing gene expression *in situ*. Environ Microbiol 16:3730–3742
Luo Z-B, Janz D, Jiang X, Gobel C, Wildhagen H, Tan Y, Rennenberg H, Feussner I, Polle A (2009) Upgrading root physiology for stress tolerance by ectomycorrhizas: insights from metabolite and transcriptional profiling into reprogramming for stress anticipation. Plant Physiol 151:1902–1917
Maillet F, Poinsot V, André O, Puech-Pagés V, Haouy A, Gueunier M, Cromer L, Giraudet D, Formey D, Niebel A, Martinez EA, Driguez H, Bécard G, Dénarié J (2011) Fungal lipochitooligosaccharide symbiotic signals in arbuscular mycorrhiza. Nature 469:58–64
Marmeisse R, Nehls U, Opik M, Selosse MA, Pringle A (2013) Bridging mycorrhizal genomics, metage-

nomics and forest ecology. New Phytol 198 (2):343–346
Marschner H, Dell B (1994) Nutrient uptake in mycorrhizal symbiosis. Plant Soil 159:89–102
Martin F (2011) Unearthing the truffle genome. New Phytol 189:645–646
Martin F, Aerts A, Ahrén D, Brun A, Danchin EGJ, Duchaussoy F, Gibon J, Kohler A, Lindquist E, Pereda V, Salamov A, Shapiro HJ, Wuyts J, Blaudez D, Buée M, Brokstein P, Canbäck B, Cohen D, Courty PE, Coutinho PM, Delaruelle C, Detter JC, Deveau A, DiFazio S, Duplessis S, Fraissinet-Tachet L, Lucic E, Frey-Klett P, Fourrey C, Feussner I, Gay G, Grimwood J, Hoegger PJ, Jain P, Kilaru S, Labbé J, Lin YC, Legué V, Le Tacon F, Marmeisse R, Melayah D, Montanini B, Murat M, Nehls U, Niculita-Hirzel H, Secq MPO, Peter M, Quesneville H, Rajashekar B, Reich M, Rouhier N, Schmutz J, Yin T, Chalot M, Henrissat B, Kües U, Lucas S, Van De Peer Y, Podila GK, Polle A, Pukkila PJ, Richardson PM, Rouzé P, Sanders IR, Stajich JE, Tunlid A, Tuskan G, Grigoriev IV (2008) The genome of *Laccaria* bicolor provides insights into mycorrhizal symbiosis. Nature 452: 88–92
Martin F, Kohler A, Murat C, Balestrini R, Coutinho PM, Jaillon O, Montanini B, Morin E, Noel B, Percudani R, Porcel B, Rubini A, Amicucci A, Amselem J, Anthouard V, Arcioni S, Artiguenave F, Aury JM, Ballario P, Bolchi A, Brenna A, Brun A, Buée M, Cantarel B, Chevalier G, Couloux A, Da Silva C, Denoeud F, Duplessis S, Ghignone S, Hilselberger B, Iotti M, Marçais B, Mello A, Miranda M, Pacioni G, Quesneville H, Riccioni C, Ruotolo R, Splivallo R, Stocchi V, Tisserant E, Viscomi AR, Zambonelli A, Zampieri E, Henrissat B, Lebrun MH, Paolocci F, Bonfante P, Ottonello S, Wincker P (2010) Périgord black truffle genome uncovers evolutionary origins and mechanisms of symbiosis. Nature 464:1033–1038
Martin F, Kohler A, Murat C, Veneault-Fourrey C, Hibbett DS (2016) Unearthing the roots of ectomycorrhizal symbioses. Nat Rev Microbiol 14: 760–773
Nehls U, Béguiristain T, Ditengou F, Lapeyrie F, Martin F (1998) The expression of a symbiosis-regulated gene in eucalypt roots is regulated by auxins and hypaphorine, the tryptophan betaine of the ectomycorrhizal basidiomycete *Pisolithus tinctorius*. Planta 207:296–302
Nibau C, Gibbs DJ, Coates JC (2008) Branching out in new directions: the control of root architecture by lateral root formation. New Phytol 179:595–614
Niemi K, Vuorinen T, Ernstsen A, Häggman H (2002) Ectomycorrhizal fungi and exogenous auxins influence root and mycorrhiza formation of Scots pine hypocotyl cuttings in vitro. Tree Physiol 22:1231–1239
Parniske M (2008) Arbuscular mycorrhiza: the mother of plant root endosymbioses. Nat Rev Microbiol 6:763–775
Pauwels L, Goossens A (2011) The JAZ proteins: a crucial interface in the jasmonate signaling cascade. Plant Cell 23:3089–3100
Pellegrin C, Morin E, Martin FM, Veneault-Fourrey C (2015) Comparative analysis of secretomes from ectomycorrhizal fungi with an emphasis on small-secreted proteins. Front Microbiol 6:1278. https://doi.org/10.3389/fmicb.2015.01278
Pellegrin C, Daguerre Y, Ruytinx J, Guinet F, Kemppainen M, Plourde MB, Hecker A, Morin E, Pardo AG, Germain H, Martin FM, Veneault-Fourrey C (2017) *Laccaria bicolor* MiSSP8 is a small-secreted protein decisive for the establishment of the ectomycorrhizal symbiosis. bioRxiv. https://doi.org/10.1101/218131
Pickles BJ, Simard SW (2016) Mycorrhizal networks and forest resilience to drought. In: Mycorrhizal mediation of soil: fertility, structure, and carbon storage. Elsevier, pp 319–339
Plett JM, Kemppainen M, Kale SD, Kohler A, Legué V, Brun A, Tyler BM, Pardo AG, Martin F (2011) A secreted effector protein of *Laccaria bicolor* is required for symbiosis development. Curr Biol 21:1197–1203
Plett JM, Daguerre Y, Wittulsky S, Vayssieres A, Deveau A, Melton SJ, Kohler A, Morrell-Falvey JL, Brun A, Veneault-Fourrey C, Martin F (2014a) Effector MiSSP7 of the mutualistic fungus *Laccaria bicolor* stabilizes the Populus JAZ6 protein and represses jasmonic acid (JA) responsive genes. Proc Natl Acad Sci 111:8299–8304
Plett JM, Khachane A, Ouassou M, Sundberg B, Kohler A, Martin F (2014b) Ethylene and jasmonic acid act as negative modulators during mutualistic symbiosis between *Laccaria bicolor* and *Populus* roots. New Phytol 202:270–286
Plett JM, Tisserant E, Brun A, Morin E, Grigoriev IV, Kuo A, Martin F, Kohler A (2015) The mutualist Laccaria bicolor expresses a core gene regulon during the colonization of diverse host plants and a variable regulon to counteract host-specific defenses. Mol Plant-Microbe Interact 28:261–273
Pozo MJ, López-Ráez JA, Azcón-Aguilar C, García-Garrido JM (2015) Phytohormones as integrators of environmental signals in the regulation of mycorrhizal symbioses. New Phytol 205:1431–1436
Regvar M, Gogala N, Žnidaršič N (1997) Jasmonic acid affects mycorrhization of spruce seedlings with *Laccaria laccata*. Trees – Struct Funct 11:511–514
Rincón A, Gérard J, Dexheimer J, Le Tacon F (2001) Effect of an auxin transport inhibitor on aggregation and attachment processes during ectomycorrhiza formation between *Laccaria bicolor* S238N and *Picea abies*. Can J Bot 79:1152–1160

Rincon CA, Raper CD, Patterson RP (2003) Genotypic differences in root anatomy affecting water movement through roots of soybean. Int J Plant Sci 164:543–551

Rovenich H, Boshoven JC, Thomma BP (2014) Filamentous pathogen effector functions: of pathogens, hosts and microbiomes. Curr Opin Plant Biol 20:96–103. https://doi.org/10.1016/j.pbi.2014.05.001

Rupp LA, Mudge KW (1985) Ethephon and auxin induce mycorrhiza-like changes in the morphology of root organ cultures of Mugo pine. Physiol Plant 64:316–322

Ryberg M, Matheny PB (2012) Asynchronous origins of ectomycorrhizal clades of Agaricales. Proc R Soc B Biol Sci 279:2003–2011

Sammer D, Krause K, Gue M, Wagner K, Kothe E (2016) Hydrophobins in the life cycle of the ectomycorrhizal basidiomycete *Tricholoma vaccinum*. PLoS One 11512:e0167773

Schmidt R, Etalo DW, Jager V, Gerards S, Zweers H, Boer W, Garbeva P (2016) Microbial small talk: volatiles in fungal-bacterial interactions. Front Microbiol 6:1495. https://doi.org/10.3389/fmicb.2015.01495

Sebastiana M, Vieira B, Lino-Neto T, Monteiro F, Figueiredo A, Sousa L, Pais MS, Tavares R, Paulo OS (2014) Oak root response to ectomycorrhizal symbiosis establishment: RNA-seq derived transcript identification and expression profiling. PLoS One 9(5):e98376

Sedjo R, Sohngen B (2012) Carbon sequestration in forests and soils. Annu Rev Resour Econ 4:127–144

Shen Q, Liu Y, Naqvi NI (2018) Fungal effectors at the crossroads of phytohormone signaling. Curr Opin Microbiol 46:1–6

Sikes BA, Cottenie K, Klironomos JN (2009) Plant and fungal identity determines pathogen protection of plant roots by arbuscular mycorrhizas. J Ecol 97:1274–1280

Simard SW, Beiler KJ, Bingham MA, Deslippe JR, Philip LJ, Teste FP (2012) Mycorrhizal networks: mechanisms, ecology and modeling. Fungal Biol Rev 26:39–60

Splivallo R, Fischer U, Gobel C, Feussner I, Karlovsky P (2009) Truffles regulate plant root morphogenesis via the production of auxin and ethylene. Plant Physiol 150:2018–2029

Steinkellner S, Lendzemo V, Langer I, Schweiger P, Khaosaad T, Toussaint JP, Vierheilig H (2007) Flavonoids and strigolactones in root exudates as signals in symbiotic and pathogenic plant-fungus interactions. Molecules 12:1290–1306

Sun J, Xu Y, Ye S, Jiang H, Chen Q, Liu F, Zhou W, Chen R, Li X, Tietz O, Wu X, Cohen JD, Palme K, Li C (2009) Arabidopsis ASA1 is important for jasmonate-mediated regulation of auxin biosynthesis and transport during lateral root formation. Plant Cell 21:1495–1511

Swarup R, Perry P, Hagenbeek D, Van Der Straeten D, Beemster GTS, Sandberg G, Bhalerao R, Ljung K, Bennett MJ (2007) Ethylene upregulates auxin biosynthesis in Arabidopsis seedlings to enhance inhibition of root cell elongation. Plant Cell 19:2186–2196

Tagu D, Kottke I, Martin F (1998) Hydrophobins in ectomycorrhizal symbiosis: hypothesis. Symbiosis 25:5–18

Tarkka MT, Herrmann S, Wubet T, Feldhahn L, Recht S, Kurth F, Mailänder S, Bönn M, Neef M, Angay O, Bacht M, Graf M, Maboreke H, Fleischmann F, Grams TEE, Ruess L, Schädler M, Brandl R, Scheu S, Schrey SD, Grosse I, Buscot F (2013) OakContigDF1591, a reference library for studying differential gene expression in *Quercus robur* during controlled biotic interactions: use for quantitative transcriptomic profiling of oak roots in ectomycorrhizal symbiosis. New Phytol 199: 529–540

Teale WD, Paponov I, Palme K (2006) Auxin in action: signalling, transport and the control of plant growth and development. Nat Rev Mol Cell Biol 7:847–859

Tsuzuki S, Handa Y, Takeda N, Kawaguchi M (2016) Strigolactone-induced putative secreted protein 1 is required for the establishment of symbiosis by the arbuscular mycorrhizal fungus *Rhizophagus irregularis*. Mol Plant-Microbe Interact 29: 277–286

van der Heijden MG, Martin FM, Selosse M-A, Sanders IR (2015) Mycorrhizal ecology and evolution: the past, the present, and the future. New Phytol 205:1406–1423

Vayssières A, Pěnčík A, Felten J, Kohler A, Ljung K, Martin F, Legué V (2015) Development of the poplar-*Laccaria bicolor* ectomycorrhiza modifies root auxin metabolism, signaling, and response. Plant Physiol 169:890–902

Vincent D, Kohler A, Claverol S, Solier E, Joets J, Gibon J, Lebrun MH, Plomion C, Martin F (2012) Secretome of the free-living mycelium from the ectomycorrhizal basidiomycete *Laccaria bicolor*. J Proteome Res 11:157–171

Wager A, Browse J (2012) Social network: JAZ protein interactions expand our knowledge of jasmonate signaling. Front Plant Sci 3:41. https://doi.org/10.3389/fpls.2012.00041

Wawra S, Fesel P, Widmer H, Timm M, Seibel J, Leson L, Kesseler L, Nostadt R, Hilbert M, Langen G, Zuccaro A (2016) The fungal-specific β-glucan-binding lectin FGB1 alters cell-wall composition and suppresses glucan-triggered immunity in plants. Nat Commun 7:13188. https://doi.org/10.1038/ncomms13188

Weiss M, Mikolajewski S, Peipp H, Schmitt U, Schmidt J, Wray V, Strack D (1997) Tissue-specific and development-dependent accumulation of phenylpropanoids in larch mycorrhizas. Plant Physiol 114:15–27

Win J, Chaparro-Garcia A, Belhaj K, Saunders DGO, Yoshida K, Dong S, Schoenack S, Zipfel C, Robatzek S, Hogenhout SA, Kamoun S (2012) Effector biology of plant-associated organisms: concepts

and perspectives. Cold Spring Harbor Laboratory Press. https://doi.org/10.1101/sqb.2012.77.015933
Woyke T, Rubin EM (2014) Searching for new branches on the tree of life. Science 80-346:698–699
Zhang F, Anasontzis GE, Labourel A, Champion C, Haon M, Kemppainen M, Commun C, Deveau A, Pardo A, Veneault-Fourrey C, Kohler A, Rosso M-N, Henrissat B, Berrin J-G, Martin F (2018) The ectomycorrhizal basidiomycete Laccaria bicolor releases a secreted endoglucanase that plays a key role in symbiosis development. New Phytol (in press)
Zhao Y (2010) Auxin biosynthesis and its role in plant development. Annu Rev Plant Biol 61:49–64

Calcium Cation Cycling and Signaling Pathways in Fungi

YUANWEI ZHANG[1], HECHUN JIANG[1], LING LU[1]

CONTENTS

I. Introduction

Ca^{2+} is chosen by all organisms as the second intracellular messenger based on its own physical and chemical properties (Carafoli et al. 2001; Haiech and Moreau 2011). Different from other second intracellular messengers, such as the cyclic adenosine monophosphate (cAMP), the inositol 1,4,5-trisphosphate (IP3), and the diacylglycerol (DAG), Ca^{2+} cannot be degraded to terminate its carried signals (Chakraborty and Hasan 2012; Hartwig et al. 2014; Hohendanner et al. 2014; Ueda et al. 2014). In contrast, Ca^{2+} ions are pumped out of the cytoplasm or conserved in the intracellular Ca^{2+} stores via a series of Ca^{2+} transporters or Ca^{2+}/anion exchangers localized in the plasma membrane or the intracellular organelle membrane, respectively. In eukaryotic cells, the vacuole, the lysosome, the endoplasmic reticulum (ER), the Golgi apparatus (GA), the mitochondria, and even the nucleus envelope are employed to function as the intracellular Ca^{2+} reservoirs (Dunn et al. 1994; Ermak and Davies 2002; Van Baelen et al. 2004; Michelangeli et al. 2005; Lustoza et al. 2011; Takeuchi et al. 2015). Upon stimulation, Ca^{2+} ions are rapidly released from the intracellular Ca^{2+} stores, resulting in the Ca^{2+} concentration increase in the cytoplasm (Cui et al. 2009). Depletion of Ca^{2+} from the intracellular Ca^{2+} stores triggers the influx of Ca^{2+} from the extracellular environments, which further increase cytosolic Ca^{2+} concentrations (Locke et al. 2000; Liu et al. 2006). These Ca^{2+} ions are caught by Ca^{2+} sensors such as calmodulin (CaM) to activate itself and subsequently the downstream enzymes such as calcineurin. The transcription factor Crz1 is consequently dephosphorylated by calcineurin and translocated from the cytoplasm into the nucleus to initiate special transcriptions in order to deal with the abovementioned stimulations (Matheos et al. 1997; Onyewu et al. 2004). Finally, the Ca^{2+} is pumped out of the cytoplasm or into the intracellular Ca^{2+} reservoirs to set the stage for the next cycle. This Ca^{2+} signaling pathway is highly conserved from prokaryotes to eukaryotes, but there are different editions in various kinds of cell types (Dominguez 2004; Tuteja and Mahajan 2007; Luan 2009; Hashimoto and Kudla 2011; Batistic and Kudla 2012). In fungal kingdom, knowledge on Ca^{2+} regulation is primarily obtained from the model fungus *Saccharomyces cerevisiae*, in spite of accumulation of parallel advances in other fungi. Here, we outline the participators or potential players in the

Yuanwei Zhang and Hechun Jiang have contributed equally to this study.

[1] Jiangsu Key Laboratory for Microbes and Functional Genomics, Jiangsu Engineering and Technology Research Center for Microbiology, College of Life Sciences, Nanjing Normal University, Nanjing, China; e-mail: linglu@njnu.edu.cn

Biology of the Fungal Cell, 3rd Edition
The Mycota VIII
D. Hoffmeister, M. Gressler (Eds.)

Ca^{2+} regulation network, focusing on Ca^{2+} circling and signaling mechanisms.

II. The Vacuole and Ca^{2+} Cycling Pathways Against the Cytoplasm

Similar to plants but different from animals, the vacuole is the principal intracellular **Ca^{2+} reservoir in fungi** (Pittman 2011). *S. cerevisiae* has one to several vacuoles in which 90% of the intracellular Ca^{2+} ions are stored (Cunningham and Fink 1994; Tanida et al. 1995). In *Neurospora crassa*, there are different types of vacuoles with distinct morphogenesis and composition, and it is not clear now whether they are just the intermediates prior to maturation or already mature, final vacuoles (Bowman et al. 2009; Cunningham 2011). Besides playing as the intracellular Ca^{2+} stores, fungal vacuoles also participate in toxic substance sequestration, osmotic pressure regulation, and pH value accommodation (Wada and Anraku 1994; Zhou et al. 2003; Moreno and Docampo 2009). Oxalate, phosphate, and organic acids inside the vacuole can chelate a bulk of Ca^{2+} ions, and only a small proportion of Ca^{2+} is ready for release in response to different stimulations (Moreno and Docampo 2009; Docampo and Moreno 2011). For example, in *S. cerevisiae* cultured in standard conditions, the free Ca^{2+} concentration in the vacuole is about 30 μM, while the total Ca^{2+} concentration is about at the level of 3 mM, indicating 99% of the Ca^{2+} has been immobilized in the vacuolar lumen (Cunningham 2011).

Membrane fusion of the vacuole with the secretory vesicles derived from the GA can contribute to Ca^{2+} uptake of the vacuole (Cunningham 2011), but two other Ca^{2+} transporting mechanisms play main roles in providing the vacuole with Ca^{2+}: a primary energized **Ca^{2+}-ATPase pathway** mediated by the plasma membrane Ca^{2+} ATPase (PMC) and a secondary energized Ca^{2+} exchanger pathway carried out by the Ca^{2+}/H^{+} exchanger (Cunningham and Fink 1996; Miseta et al. 1999b; Samarao et al. 2009; Li et al. 2011; Pittman 2011). Ca^{2+}-ATPases belong to the P-type ATPase superfamily characterized by the formation of a phosphorylated intermediate during the transportation cycle (Okamura et al. 2003; Bublitz et al. 2010). This superfamily is divided into five families based on sequence similarities, phylogenetic analyses, and especially substrate specificities, with Ca^{2+}-ATPase being grouped as type II P-type ATPase (Okamura et al. 2003). In contrast to the high-affinity property of the Ca^{2+}-ATPase, Ca^{2+}/H^{+} exchanger has the characteristics of low affinity but accompanied by high capacity, which can transport Ca^{2+} to the vacuole with high efficiency soon after a burst of cytosolic Ca^{2+} elevation (Pittman 2011).

S. cerevisiae **Pmc1** was the first Ca^{2+} ATPase identified and cloned from all organisms, which displays 40% identical to the mammalian plasma membrane Ca^{2+} ATPases (PMCAs). In *S. cerevisiae*, *pmc1* mutants sequestrate Ca^{2+} in the vacuole at less than 20% of the wild type in the standard medium (Cunningham and Fink 1994), and the mutants demonstrate sensitivity to high concentrations of environmental Ca^{2+} (Cunningham and Fink 1996), suggesting that Pmc1 probably plays predominant roles in Ca^{2+} sequestration in the vacuole. In the opportunistic pathogen *Cryptococcus neoformans*, deletion of *pmc1* impairs blood-brain barrier transmigration and alters the global gene expression profile, including *ure1* which encodes the virulence factor urease (Squizani et al. 2018). In *S. cerevisiae*, Pmc1 almost settles exclusively in the vacuolar membrane, whereas in *N. crassa*, the Pmc1 counterparts NCA-2 and NCA-3 localize in the vacuolar membrane as well as in the plasma membrane (Bowman et al. 2011). Mammalian PMCAs are solely positioned in the plasma membrane and possess an autoinhibition domain at the C-terminus, and binding of the Ca^{2+} sensor calmodulin with this regulatory domain can release the activities of PMCAs (Brini and Carafoli 2009). In *S. cerevisiae*, Pmc1 has no C-terminal inhibitory region, and the regulation takes place at the transcription level (Cunningham and Fink 1996; Matheos et al. 1997). Hence, the expression level of *pmc1* mRNA is upregulated as a response to the elevated cytosolic Ca^{2+} level.

Vcx1 (vacuolar calcium exchanger), as the first Ca^{2+}/H^{+} exchanger identified in eukaryotes, was originally discovered to be important for *S. cerevisiae pmc1* mutants to survive in high levels of Ca^{2+} (Cunningham and Fink 1996; Miseta et al. 1999b). Vcx1 had been demonstrated to play a primary role in transporting Ca^{2+} to the vacuole in *S. cerevisiae*. A similar phenomenon is also observed in *C. neoformans*, which showed that *C. neoformans vcx1* null mutants had an increased intracellular calcium concentration (Kmetzsch et al. 2010). However, deletion of *S. cerevisiae vcx1* does not lead to obvious decrease of vacuolar Ca^{2+} concentrations, and the mutant does not exhibit sensitivity to high environmental Ca^{2+} (Cunningham and Fink 1996; Miseta et al. 1999b), which is in sharp contrast to the phenotypes of *pmc1* mutant. In fact, Ca^{2+} transport into the vacuole is mainly mediated by Pmc1 and to a lower extent by Vcx1 resulting from the regulation of calcineurin, the calmodulin-dependent serine/threonine phosphatase. Calcineurin increases the expression levels of Pmc1 and inhibits the activity of Vcx1 simultaneously (Cunningham and Fink 1996; Kmetzsch et al. 2010). The exact reason for this phenomenon remains unexplored, but one can hypothesize that cells avoid Ca^{2+} overload by not activating both Ca^{2+} transporters simultaneously.

Ca^{2+} can rapidly enter the vacuole via Vcx1, with the stoichiometry of 1 Ca^{2+} ion per 2 or 3 H^+ ions (Dunn et al. 1994), immediately after the cytosolic Ca^{2+} concentrations are increased and prior to the complete activation of calcineurin. The pH gradient across the membrane of the vacuole is the prerequisite for Vcx1 to function normally, which is administered by the H^+ ATPase (V-type ATPase) in the vacuolar membrane (Kakinuma et al. 1981). A sudden loss of vacuole acidity might not only prevent Ca^{2+} influx into the vacuole but also oppositely induce the Ca^{2+} efflux from the vacuole via Vcx1 (Forster and Kane 2000). When the gene encoding the H^+ ATPase was mutated, yeast cannot tolerate high concentration of Ca^{2+}, further confirming that acidification of vacuole is indispensable for Ca^{2+} efflux into lumen (Forster and Kane 2000).

Ca^{2+} could possibly be released from the vacuole via Vcx1 in response to special conditions (Cunningham 2011); however, the vacuole is equipped with at least one Ca^{2+} channel specially responsible for Ca^{2+} release. **Yvc1 (yeast vacuolar conductance)** is a homologue of mammalian Ca^{2+} channels from TRPC (transient receptor potential canonical) family (Palmer et al. 2001; Cunningham 2011; Yu et al. 2014), which can convey Ca^{2+} and some other ions such as K^+ and Na^+ across the membrane (Chang et al. 2010). To date, three kinds of Ca^{2+} channels have been identified: voltage-gated channels, stretch-activated channels, and ligand-gated channels (Carafoli et al. 2001). Studies on Yvc1 from isolated *S. cerevisiae* vacuoles indicate that it can be activated by membrane stretch, suggesting that Yvc1 is a mechanosensitive (stretch-activated) ion channel (Zhou et al. 2003; Loukin et al. 2008; Chang et al. 2010). Another evidence suggests that Yvc1 might also be a ligand-activated channel, because phosphatidylinositol-3,5 bisphosphate (PIP_2) and Ca^{2+} itself can trigger Ca^{2+} release from the vacuole via binding to Yvc1 (Dong et al. 2010). Ca^{2+}-induced Ca^{2+} release is common with the Ca^{2+} efflux from sarcoplasmic/endoplasmic reticulum of mammalian cells (Hernandez-Cruz et al. 1997; Ji et al. 2006). Considering the fact that the vacuole is the primary Ca^{2+} detoxifying organelle in the cell, this mechanism might render Pmc1 and Vcx1 a huge mission to sequester Ca^{2+} from the cytoplasm. At least one other Ca^{2+} channel, rather than Yvc1, is situated in fungal vacuolar membrane, because the ligand IP_3 is able to trigger Ca^{2+} release from the vacuole via the IP_3 receptor Ca^{2+} channel (IP_3R) (Calvert and Sanders 1995; Silverman-Gavrila and Lew 2002), but the isolated Yvc1 does not respond to IP_3 stimulation under numbers of patch-clamp conditions in *S. cerevisiae* (Palmer et al. 2001). This IP_3-mediated Ca^{2+} channel induced by IP_3 has not been identified to date and needs further exploitation.

Ca^{2+} Circulation Between the Golgi and the Cytoplasm

The **Golgi apparatus (or Golgi complex, GA)** was first discovered in 1898 by the Italian physician Camillo Golgi (Bentivoglio et al. 2011; Li et al. 2013). During the next several decades, debates still exit focusing on the existence of this organelle in the cell. Now, it is clear that the GA is an independent intracellular organelle, and it is constituted of stacks of flattened cisternae with polarity, which are primarily organized into three compartments: the cis-Golgi network (CGN), the medial-Golgi, and the trans-Golgi network (TGN) (Pizzo et al. 2011; Day et al. 2013). Although the overall morphogenesis of the GA might vary slightly, for example, the cisternae seem relatively disperse in *S. cerevisiae* whereas compact in *Schizosaccharomyces pombe*, their functions are believed to be similar (Suda and Nakano 2012).

The GA is the processing factory and sorting center in the cell, guaranteeing cargo proteins to be modified, packaged, and sent to their final destinations correctly (Nakano 2004; Pizzo et al. 2011). Some processing enzymes in the medial-Golgi or TGN utilized for glycosylation need Ca^{2+} as the cofactor, while correct aggregation of secretory proteins in the TGN relies on the Ca^{2+} concentration in lumen, indicating the GA might be an intracellular Ca^{2+} reservoir (Micaroni and Mironov 2010; Pizzo et al. 2011; von Blume et al. 2011). The GA contains Ca^{2+} ions at the high concentration, ranging from 10^{-4} to 10^{-3} M in mammalian cells (Li et al. 2013), which is distributed unevenly in different subcompartments from CGN to TGN. In the membrane of the GA, there exist the Ca^{2+} ATPase(s), responsible for transporting Ca^{2+} from cytoplasm to the GA lumen and the Ca^{2+} channel(s), releasing Ca^{2+} from the GA lumen to the cytoplasm (Van Baelen et al. 2004; Pizzo et al. 2011; Li et al. 2013).

Pmr1 (plasma membrane ATPase related) was identified in *S. cerevisiae* (Rudolph et al. 1989) as the first Ca^{2+} ATPase found to be situated in the GA, especially the medial-Golgi, TGN, and secretory vesicles derived from TGN (Antebi and Fink 1992; Durr et al. 1998). Further homologues of Pmr1 have been identified from a variety of organisms, such as other fungi, worms, insects, mammals, and even bacteria (Yagodin et al. 1999; Van Baelen et al. 2001; Raeymaekers et al. 2002; Bates et al. 2005; Cho et al. 2005; Buttner et al. 2013). Pmr1 belongs to the type II P-type ATPase, but it is distinguished from another member of the type II P-type ATPase, i.e., SERCA (sarcoplasmic/endoplasmic reticulum Ca^{2+} ATPase) in at least three aspects: (1) Pmr1 translocates one Ca^{2+} ion at the expense of one molecule of ATP, whereas SERCA can simultaneously transport two Ca^{2+} per ATP molecule (Wuytack et al. 2003); (2) Pmr1 is able to transfer not only Ca^{2+} but also Mn^{2+} (Ton et al. 2002); and (3) ion transportation of Pmr1 is thapsigargin-insensitive, whereas Ca^{2+} transport action of SERCA is sensitive to thapsigargin (Wuytack et al. 2003). The possible reason for the difference in transport efficiency between Pmr1 and SERCA is that Pmr1 has only one Ca^{2+}-binding site, while SERCA has two (Wuytack et al. 2003). The fact that the unique ion-binding site in Pmr1 can be occupied by either Ca^{2+} or Mn^{2+} is probably due to the relatively low substrate specificity of this Ca^{2+}/Mn^{2+} ATPase (He and Hu 2012).

S. cerevisiae lacks Ca^{2+} ATPase(s) in the plasma membrane and ER membrane (i.e., PMCA and SERCA), and Vcx1 function is inhibited by calcineurin activation. These may suggest that Pmr1 and Pmc1 play the predominant role in maintaining the cytosolic homeostasis. Deletion of *pmr1* in *S. cerevisiae* leads to 16-fold increase of cytosolic Ca^{2+} levels relative to wild type (Ton et al. 2002). In the *S. cerevisiae pmr1* mutants, the expression level of *pmc1* rises about fivefold, and Pmc1-splicing variants might be localized in the GA, as the compensatory mechanisms (Marchi et al. 1999; Miseta et al. 1999a). Double deletion of *pmr1* and *pmc1* in *S. cerevisiae* is lethal in standard medium. However, repression of the calcineurin activity, which abolishes the inhibition of Vcx1, will recover the viability (Kellermayer et al. 2003). Different from the case in *S. cerevisiae*, double deletion of *pmr1* and *pmc1* homologues in *Aspergillus nidulans* is not lethal, which probably results from Ca^{2+} detoxification carried out by the Ca^{2+} ATPase(s) in the plasma membrane or in ER membrane (Jiang et al. 2014a).

The GA, possessing about 5% of total intracellular Ca^{2+} in LLC-PK1 cells (Pizzo et al. 2011), is suggested to be an exchangeable Ca^{2+} store; therefore, at least one channel is needed to mediate Ca^{2+} release from the lumen to the cytoplasm. IP_3 receptor (IP_3R) is a Ca^{2+} channel situated primarily in the sarcoplasmic/endoplasmic reticulum membrane, mediating Ca^{2+} influx from matrix to cytoplasm (Krizanova and Ondrias 2003; Bezprozvanny 2005). Recently, IP_3R is found to reside in the GA as well, because the intracellular second messenger IP_3 is able to trigger Ca^{2+} mobilization from the GA, especially the CGN subcompartment (Li et al. 2013). Ryanodine receptors (RyR) is sarcoplasmic/endoplasmic reticulum-localized Ca^{2+} channel that is responsible for the release of Ca^{2+} from intracellular stores (Lanner et al. 2010). It has been demonstrated that RyR may participate in Ca^{2+} release from the GA in rat sympathetic neurons and exist in neonatal cardiac myocytes as well (Cifuentes et al. 2001; Lissandron et al. 2010). Therefore, it is possible that RyR might also exist in GA membrane. However, no IP_3R and RyR Ca^{2+} channels have been reported in the fungal kingdom.

III. Ca^{2+} Circuit from and to the Endoplasmic Reticulum

Compared to the vacuole and the GA, **ER (endoplasmic reticulum)** is less important in sequestering Ca^{2+} from the cytoplasm in fungi. Nevertheless, Ca^{2+} is maintained in the fungal ER lumen, e.g., *S. cerevisiae* has about 10 μM free Ca^{2+} in ER matrix (Ton et al. 2002; Li et al. 2013), which is crucial to the organelle's function, especially in the perspective of protein modification and folding, because a variety of ER-resident enzymes are Ca^{2+}-dependent (Uemura et al. 2007; Losev et al. 2008). For example, Calreticulin is the molecule chaperon in ER contributing to correct protein folding, which is also a Ca^{2+}-binding protein with high capacity (Parlati et al. 1995; Groenendyk et al. 2004; Brunner et al. 2012). There are many other chaperons in the ER lumen, such as Calnexin, endoplasmin, PDI (protein disulfide isomerase), ERp (ER protein), and BiP (immunoglobulin-binding protein) (Shnyder et al. 2008). They corporately support correct protein folding in a Ca^{2+}-dependent matter [for details, please refer to (Michalak et al. 2002; Groenendyk et al. 2004; Coe and Michalak 2009)].

In mammalian cells, the Ca^{2+} ATPase replenishing ER with Ca^{2+} is dependent on SERCA (sarcoplasmic/endoplasmic reticulum calcium ATPase), which was first discovered in 1962 when sarcoplasmic reticulum fractions were shown to accumulate Ca^{2+} at the expenses of ATP hydrolysis (Carafoli and Brini 2000). SERCA belongs to type II P-type ATPase, translocating Ca^{2+} and/or Mn^{2+} across the membrane at the expense of ATP hydrolysis. In *S. cerevisiae*, no SERCA homologues have been identified from the ER membrane, and partial ER luminal Ca^{2+} is contributed by Pmr1 located mainly in the GA. However, type V P-type ATPase Cod1/Spf1 localized in the ER membrane can supplement ER with Ca^{2+}, although the substrate transfer and the in vivo function of this P-type ATPase are not clear yet (Cronin et al. 2002). *S. pombe* type V P-type ATPase Cta4p is proven to have Ca^{2+}-transporting activity coupled with ATPase activity, and a *S. pombe* mutant lacking *cta4* has a sixfold excess of cytosolic Ca^{2+} concentration when compared to wild type (Lustoza et al. 2011). On the contrary, there is SERCA homologue in the ER membrane in *N. crassa*, namely, NCA-1, which has been proved to be essential in maintaining the calcium level (Bowman et al. 2011).

IV. Ca^{2+} Exchange Between Mitochondria and the Cytoplasm

The **mitochondrion** is an essential intracellular organelle for energy production, because the tricarboxylic acid (TCA) cycle and the oxidative phosphorylation coupled by the electron transfer reactions take place in the lumen or across the membrane (Pan et al. 2007; Katoh et al. 2008). Some dehydrogenases such as the pyruvate dehydrogenase, the oxoglutarate dehydrogenase, and the isocitrate dehydrogenase employed in the ATP creation procedures need Ca^{2+} at high concentrations to increase enzyme activities (Denton 2009; Santo-Domingo and Demaurex 2010). Therefore, it is reasonable to deduce that the inner Ca^{2+} concentration of the mitochondria can be regulated to satisfy with different physiological demands. In fact, it has been established for a long period of time that the mitochondria can sequester and release Ca^{2+} as a dynamic Ca^{2+} store (Ganitkevich 2003). As other intracellular Ca^{2+} sinks, only a small percentage of Ca^{2+} is freely available, and most of it is bound by the buffers in the matrix, probably membrane phospholipids and/or the inner phosphate ions (Horikawa et al. 1998; Pivovarova et al. 1999; Ganitkevich 2003).

To date, two routes mediating Ca^{2+} entering the mitochondria have been identified: a saturable **low-affinity uniporter pathway** and a saturable **rapid uptake mode (RaM)** (Gunter and Gunter 2001; Adiele et al. 2012). The **mitochondrial Ca^{2+} uniporter (MCU)** located in the inner membrane is believed to be mainly responsible for Ca^{2+} accumulation in this organelle, which is driven by the mitochondrial membrane potential established by the electron-transport chain (Santo-Domingo and Demaurex 2010; Samanta et al. 2014). MCU has a high preference for Ca^{2+} and requires a high concentration of external Ca^{2+} to open the gate. This seems to be the reason why mitochondria are always localized closely to the plasma membrane and the ER membrane (in mammalian cells), because clusters of mitochondrial Ca^{2+} channels can provide bursts of high concentrations of Ca^{2+} (Michelangeli et al. 2005; Prole and Taylor 2012; Takeuchi et al. 2015). MCU have been identified in many eukaryotic cells but are absent in *S. cerevisiae* (Kovacs-Bogdan et al. 2014), while a recent report in the fungal pathogen *Aspergillus fumigatus* demonstrated that deletion of *mcuA*, which is a putative mammalian MCU gene homolog, impaired the mitochondrial Ca^{2+} homeostasis and environmental stress adaptation (Song et al. 2016).

RaM acts as another Ca^{2+} entry pathway in the membrane of the mitochondria and probably owns more important physiological functions than MCU (Buntinas et al. 2001; Bazil and Dash 2011). Different from MCU, RaM can mediate Ca^{2+} influx in the resting state of cytosolic Ca^{2+} concentrations, supplying enough Ca^{2+} to activate the dehydrogenases using Ca^{2+} as a cofactor. The RaM-mediated Ca^{2+} influx is terminated immediately after its initiation, preventing overload of Ca^{2+} to the mitochondrial matrix (Ganitkevich 2003). A considerable increase of mitochondrial Ca^{2+} leads to the formation of the permeability transition pore (PTP) in the mitochondrial membrane, a voltage-gating nonspecific channel with numerous subunits, which can release a wide range of matrix materials out of the mitochondria into the cytoplasm, including Ca^{2+}, via PTP itself or the impaired mitochondrial membrane induced by environmental stresses (Bernardi and Di Lisa 2014; De Marchi et al. 2014).

Ca^{2+} can be released from the mitochondrion via PTP; however, Ca^{2+} efflux from the mitochondrion is primarily mediated by two other saturable pathways, a **Na^+-dependent pathway (Ca^{2+}/Na^+ exchanger)** and a **Na^+-independent pathway (Ca^{2+}/H^+ exchanger)** (Pfeiffer et al. 2001; Molinaro et al. 2013; Tsai et al. 2014). The Ca^{2+}/Na^+ exchanger transports one Ca^{2+} ion out of the mitochondria matrix and in return three Na^+ ions into the lumen (Pfeiffer et al. 2001). Therefore, Ca^{2+} release from the mitochondria via the Ca^{2+}/Na^+ exchanger is accompanied by the accumulation of the mitochondrial Na^+ ions, which are subsequently extruded out of the matrix via Na^+/H^+ exchanger for recovering the previous ion gradients (Blomeyer et al. 2013; Rueda et al. 2014). Distinct from the Ca^{2+}/Na^+ exchanger, the H^+/Ca^{2+} exchanger directly utilizes the energy derived from the H^+ potential across the membrane to drive Ca^{2+} efflux out of the mitochondria with the stoichiometry of n H^+: Ca^{2+}, $n \geq 2$ (Pfeiffer et al. 2001; Ganitkevich 2003).

V. Ca^{2+} Signaling Pathways and Its Primary Components in the Cytoplasm

In addition to sequestering cytosolic Ca^{2+} in the intracellular stores, eukaryotic cells also clear Ca^{2+} out of the cytoplasm via the **plasma membrane Ca^{2+} ATPase (PMCA)** and a Ca^{2+}/ion exchanger localized in the plasma membrane (Herchuelz et al. 2013; Roome and Empson 2013). The first PMCA was identified from erythrocytes by Schatzmann in 1966. Afterwards, these kind of Ca^{2+} pumps are found to exist in almost all other eukaryotic cells (Carafoli and Brini 2000). However, no PMCA homologues have been characterized in fungi yet. In contrast, many researches have been performed on the Ca^{2+} channels in the fungal plasma membrane. There are two Ca^{2+} entry systems responsible for extracellular Ca^{2+} influx: (1) the **high-affinity calcium channel (HACS)**, which has high affinity for Ca^{2+} and works in the environment containing Ca^{2+} at low concentration, and (2) the **low-affinity calcium channel (LACS)**, which has relatively reduced

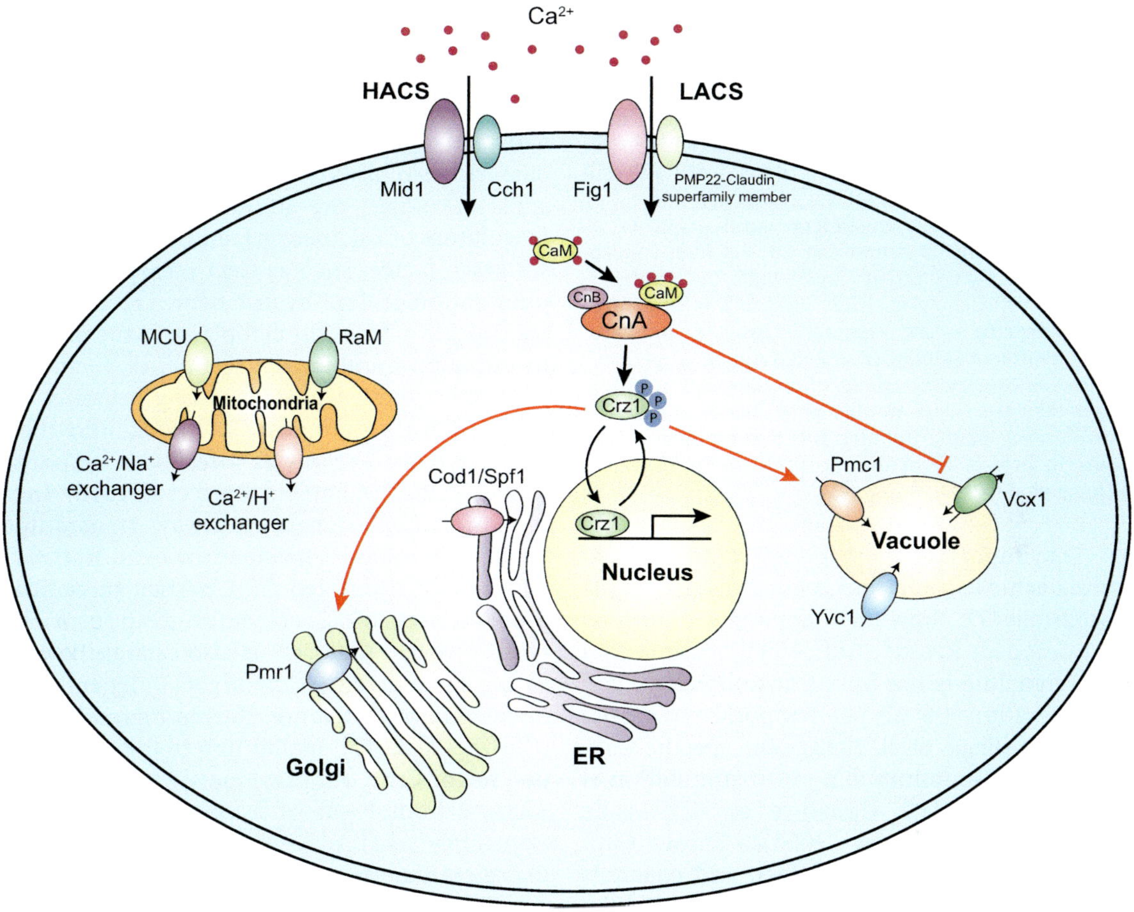

Fig. 1 Scheme model of calcium signaling in eukaryotic cells. Calcium enters through the HACS and/or LACS or is released from calcium stores, mitochondria, and vacuole. Then, Calcineurin binds to the calmodulin-Ca^{2+} complex and dephosphorylates nuclear the transcription factor Crz1 in the cytoplasm, leading to its nuclear translocation. Crz1 subsequently activates downstream genes

affinity for Ca^{2+} and works in circumstances with a higher Ca^{2+} concentration (Muller et al. 2001; Cavinder et al. 2011; Martin et al. 2011) (Fig. 1). The HACS is constituted of at least three subunits: the stretch-activated Ca^{2+} channel **Mid1 (mating-induced death)**; the **voltage-gated calcium channel Cch1 (calcium-channel homologue)**; and the **regulatory protein Ecm7 (extracellular matrix)** (Ding et al. 2013). The LACS is minimally comprised of the Fig1 protein (mating factor-induced gene) and a PMP22-Claudin superfamily member (Brand et al. 2007; Jiang et al. 2014b).

Mid1 was first identified in *S. cerevisiae* and is found to be essential for this organism to mate in presence of environmental Ca^{2+} at low concentrations (Iida et al. 1994; Kanzaki et al. 1999). During the mating cycle, the two haploid cells of opposite mating types (MATa and MATα) fuse to form a diploid zygote (Fischer et al. 1997), which is initiated by binding of pheromone (a factor or α factor secreted by MATa and MATα) to the receptor in the membrane of the other mating type. Then, a series of programmed actions are induced, such as agglutination of cells, regulation of the cell cycle, and differentiation of the cell into shmoos (differentiated cells with projects for mating), accompanied by Ca^{2+} influx from extracellular environments. *S. cerevisiae* lacking Mid1, Cch1, or both cannot complete the mating procedure and finally die upon the pheromone treatment (Iida et al. 1994; Muller et al. 2001). In *Aspergillus* species, deletion of *cchA* or *midA* causes the defects in the conidiation, hyphal polarity, and cell wall integrity in *A. nidulans* (Wang et al. 2012). In comparison, the *A. fumigatus midA* deletion mutant is hypervirulent in the immunosuppressed mice model (Jiang et al. 2014b). In addition, a coimmunoprecipitation test demonstrated that Mid1 and Cch1 in *S. cerevisiae* might assemble to form a Ca^{2+} channel (Liu et al. 2006). It is suggested that Mid1 acts as a regulatory subunit, whereas Cch1 is the catalytic subunit in the HACS; however, Mid1 expressed in COS7 or CHO cells demonstrates the ability of Ca^{2+} transfer, indicating that Mid1 can function independently (Liu et al. 2006). In rich media containing Ca^{2+} at high concentrations, the HACS activity to mediate Ca^{2+} influx is inhibited by calcineurin, while the LACS is stimulated to play the role as Ca^{2+} channel in response to pheromone treatment (Muller et al. 2003; Aguilar et al. 2007; Yang et al. 2011), although its affinity for Ca^{2+} is 16-fold lower than the HACS (Muller et al. 2001). In *A. nidulans*, lacking of the putative LACS components FigA leads to defects in hyphal growth and development (Zhang et al. 2014).

One important question regarding Ca^{2+} signaling is how signals and cellular events couple specifically, i.e., how Ca^{2+} signaling is encoded and decoded. Recent studies demonstrate that Ca^{2+} signaling is not only transmitted by Ca^{2+} concentrations but also by temporal and spatial messages (Flegg et al. 2013). The specific Ca^{2+} profile corresponding to a given stimulation is often termed Ca^{2+} signature or Ca^{2+} code (White and Broadley 2003). Those Ca^{2+} pumps, exchangers, antiporters, and channels all participate in shaping Ca^{2+} signals. In the cytoplasm, a set of EF-hand proteins function as Ca^{2+} buffers in shaping the Ca^{2+} signals. However, the majority of the EF-hand proteins acts as Ca^{2+} sensors, which deliver Ca^{2+} signals downstream via interaction with target enzymes after conformational changes (Schwaller 2010). **Calmodulin (CaM)** is the well-known representative of Ca^{2+} sensors, which has a wide range of downstream target proteins, including calcineurin, a conserved serine/threonine phosphatase composed of the catalytic subunit calcineurin-A (CnA) and the regulatory subunit calcineurin-B (CnB) (Rusnak and Mertz 2000). CnA contains four primary regions: the N-terminal catalytic domain similar with other phosphatases; CnB-binding region; CaM-binding segment; and C-terminal autoinhibitory region (Baksh and Burakoff 2000; Rusnak and Mertz 2000). CnA alone has relatively a low phosphatase activity, but association with CnB enhances its activity greatly, probably resulting from repressing the autoinhibition (Perrino et al. 1992). CnB is homologous to CaM, and they may play similar roles in CnA activation (Stemmer and Klee 1994). Calcineurin also has a broad range of downstream substrates, i.e., the ion channels, the RCN (regulators of calcineurin) family, and the transcription factor, such as Crz1 (crazy). Dephosphorylation of Crz1 by calcineurin results in its translocation from the cytoplasm to the nucleus to initiate a number of target genes, including *pmr1* and *pmc1*, whose encoding proteins are Ca^{2+} ATPases pumping Ca^{2+} back into their respective Ca^{2+} stores as a feedback response (Liu et al. 2015; Chatfield-Reed et al. 2016). In *S. cerevisiae*, Ca^{2+} signals are always transmitted by the calmodulin-calcineurin-Crz1 pathway and finally translated into particular cellular events (Cyert 2003). A recent study in *C. neoformans* revealed that Crz1 is also capable of activating some downstream targets independently of calcineurin activation during thermal stress (Chow et al. 2017). In addition to the transcription factor Crz1, our previous work showed that AkrA, a homologue of *S. cerevisiae* palmitoyl transferase Akr1, plays a vital role in maintaining calcium homeostasis in *Aspergilli*. The *akrA* deletion mutant showed significantly decreased cytosolic-free Ca^{2+} levels induced by external stresses including high extracellular Ca^{2+} stress, ER stress, and membrane stress, suggesting that posttranslational modification may also be involved in cytosolic Ca^{2+} homeostasis in *Aspergilli* (Zhang et al. 2016).

References

Adiele RC, Stevens D, Kamunde C (2012) Features of cadmium and calcium uptake and toxicity in rainbow trout (*Oncorhynchus mykiss*) mitochondria. Toxicol In Vitro 26(1):164–173

Aguilar PS, Engel A, Walter P (2007) The plasma membrane proteins Prm1 and Fig1 ascertain fidelity of membrane fusion during yeast mating. Mol Biol Cell 18(2):547–556

Antebi A, Fink GR (1992) The yeast Ca(2+)-ATPase homologue, PMR1, is required for normal Golgi function and localizes in a novel Golgi-like distribution. Mol Biol Cell 3(6):633–654

Baksh S, Burakoff SJ (2000) The role of calcineurin in lymphocyte activation. Semin Immunol 12(4): 405–415
Bates S, MacCallum DM, Bertram G, Munro CA, Hughes HB, Buurman ET, Brown AJ, Odds FC, Gow NA (2005) *Candida albicans* Pmr1p, a secretory pathway P-type Ca2+/Mn2+-ATPase, is required for glycosylation and virulence. J Biol Chem 280(24):23408–23415
Batistic O, Kudla J (2012) Analysis of calcium signaling pathways in plants. Biochim Biophys Acta 1820 (8):1283–1293
Bazil JN, Dash RK (2011) A minimal model for the mitochondrial rapid mode of Ca(2)+ uptake mechanism. PLoS One 6(6):e21324
Bentivoglio M, Jones EG, Mazzarello P, Ribak CE, Shepherd GM, Swanson LW (2011) Camillo Golgi and modern neuroscience. Brain Res Rev 66 (1–2):1–4
Bernardi P, Di Lisa F (2014) The mitochondrial permeability transition pore: molecular nature and role as a target in cardioprotection. J Mol Cell Cardiol 78:100–106
Bezprozvanny I (2005) The inositol 1,4,5-trisphosphate receptors. Cell Calcium 38(3–4):261–272
Blomeyer CA, Bazil JN, Stowe DF, Pradhan RK, Dash RK, Camara AK (2013) Dynamic buffering of mitochondrial Ca2+ during Ca2+ uptake and Na +-induced Ca2+ release. J Bioenerg Biomembr 45 (3):189–202
Bowman BJ, Draskovic M, Freitag M, Bowman EJ (2009) Structure and distribution of organelles and cellular location of calcium transporters in *Neurospora crassa*. Eukaryot Cell 8(12):1845–1855
Bowman BJ, Abreu S, Margolles-Clark E, Draskovic M, Bowman EJ (2011) Role of four calcium transport proteins, encoded by nca-1, nca-2, nca-3, and cax, in maintaining intracellular calcium levels in *Neurospora crassa*. Eukaryot Cell 10(5):654–661
Brand A, Shanks S, Duncan VM, Yang M, Mackenzie K, Gow NA (2007) Hyphal orientation of *Candida albicans* is regulated by a calcium-dependent mechanism. Curr Biol 17(4):347–352
Brini M, Carafoli E (2009) Calcium pumps in health and disease. Physiol Rev 89(4):1341–1378
Brunner JM, Plattet P, Doucey MA, Rosso L, Curie T, Montagner A, Wittek R, Vandelvelde M, Zurbriggen A, Hirling H, Desvergne B (2012) Morbillivirus glycoprotein expression induces ER stress, alters Ca2+ homeostasis and results in the release of vasostatin. PLoS One 7(3):e32803
Bublitz M, Poulsen H, Morth JP, Nissen P (2010) In and out of the cation pumps: P-type ATPase structure revisited. Curr Opin Struct Biol 20(4):431–439
Buntinas L, Gunter KK, Sparagna GC, Gunter TE (2001) The rapid mode of calcium uptake into heart mitochondria (RaM): comparison to RaM in liver mitochondria. Biochim Biophys Acta 1504(2–3): 248–261
Buttner S, Faes L, Reichelt WN, Broeskamp F, Habernig L, Benke S, Kourtis N, Ruli D, Carmona-Gutierrez D, Eisenberg T, D'Hooge P, Ghillebert R, Franssens V, Harger A, Pieber TR, Freudenberger P, Kroemer G, Sigrist SJ, Winderickx J, Callewaert G, Tavernarakis N, Madeo F (2013) The Ca2+/Mn2+ ion-pump PMR1 links elevation of cytosolic Ca (2+) levels to alpha-synuclein toxicity in Parkinson's disease models. Cell Death Differ 20 (3):465–477
Calvert CM, Sanders D (1995) Inositol trisphosphate-dependent and -independent Ca2+ mobilization pathways at the vacuolar membrane of *Candida albicans*. J Biol Chem 270(13):7272–7280
Carafoli E, Brini M (2000) Calcium pumps: structural basis for and mechanism of calcium transmembrane transport. Curr Opin Chem Biol 4(2): 152–161
Carafoli E, Santella L, Branca D, Brini M (2001) Generation, control, and processing of cellular calcium signals. Crit Rev Biochem Mol Biol 36 (2):107–260
Cavinder B, Hamam A, Lew RR, Trail F (2011) Mid1, a mechanosensitive calcium ion channel, affects growth, development, and ascospore discharge in the filamentous fungus *Gibberella zeae*. Eukaryot Cell 10(6):832–841
Chakraborty S, Hasan G (2012) IP3R, store-operated Ca2+ entry and neuronal Ca2+ homoeostasis in *Drosophila*. Biochem Soc Trans 40(1):279–281
Chang Y, Schlenstedt G, Flockerzi V, Beck A (2010) Properties of the intracellular transient receptor potential (TRP) channel in yeast, Yvc1. FEBS Lett 584(10):2028–2032
Chatfield-Reed K, Vachon L, Kwon EJ, Chua G (2016) Conserved and diverged functions of the calcineurin-activated Prz1 transcription factor in fission yeast. Genetics 202(4):1365–1375
Cho JH, Ko KM, Singaravelu G, Ahnn J (2005) *Caenorhabditis elegans* PMR1, a P-type calcium ATPase, is important for calcium/manganese homeostasis and oxidative stress response. FEBS Lett 579(3): 778–782
Chow EW, Clancey SA, Billmyre RB, Averette AF, Granek JA, Mieczkowski P, Cardenas ME, Heitman J (2017) Elucidation of the calcineurin-Crz1 stress response transcriptional network in the human fungal pathogen *Cryptococcus neoformans*. PLoS Genet 13(4):e1006667
Cifuentes F, Gonzalez CE, Fiordelisio T, Guerrero G, Lai FA, Hernandez-Cruz A (2001) A ryanodine fluorescent derivative reveals the presence of high-affinity ryanodine binding sites in the Golgi complex of rat sympathetic neurons, with possible functional roles in intracellular Ca(2+) signaling. Cell Signal 13(5):353–362
Coe H, Michalak M (2009) Calcium binding chaperones of the endoplasmic reticulum. Gen Physiol Biophys 28 Spec No Focus:F96–F103

Cronin SR, Rao R, Hampton RY (2002) Cod1p/Spf1p is a P-type ATPase involved in ER function and Ca2+ homeostasis. J Cell Biol 157(6):1017–1028

Cui J, Kaandorp JA, Ositelu OO, Beaudry V, Knight A, Nanfack YF, Cunningham KW (2009) Simulating calcium influx and free calcium concentrations in yeast. Cell Calcium 45(2):123–132

Cunningham KW (2011) Acidic calcium stores of *Saccharomyces cerevisiae*. Cell Calcium 50(2):129–138

Cunningham KW, Fink GR (1994) Calcineurin-dependent growth control in *Saccharomyces cerevisiae* mutants lacking PMC1, a homolog of plasma membrane Ca2+ ATPases. J Cell Biol 124 (3):351–363

Cunningham KW, Fink GR (1996) Calcineurin inhibits VCX1-dependent H+/Ca2+ exchange and induces Ca2+ ATPases in *Saccharomyces cerevisiae*. Mol Cell Biol 16(5):2226–2237

Cyert MS (2003) Calcineurin signaling in *Saccharomyces cerevisiae*: how yeast go crazy in response to stress. Biochem Biophys Res Commun 311 (4):1143–1150

Day KJ, Staehelin LA, Glick BS (2013) A three-stage model of Golgi structure and function. Histochem Cell Biol 140(3):239–249

De Marchi E, Bonora M, Giorgi C, Pinton P (2014) The mitochondrial permeability transition pore is a dispensable element for mitochondrial calcium efflux. Cell Calcium 56(1):1–13

Denton RM (2009) Regulation of mitochondrial dehydrogenases by calcium ions. Biochim Biophys Acta 1787(11):1309–1316

Ding X, Yu Q, Xu N, Wang Y, Cheng X, Qian K, Zhao Q, Zhang B, Xing L, Li M (2013) Ecm7, a regulator of HACS, functions in calcium homeostasis maintenance, oxidative stress response and hyphal development in *Candida albicans*. Fungal Genet Biol 57:23–32

Docampo R, Moreno SN (2011) Acidocalcisomes. Cell Calcium 50(2):113–119

Dominguez DC (2004) Calcium signalling in bacteria. Mol Microbiol 54(2):291–297

Dong XP, Shen D, Wang X, Dawson T, Li X, Zhang Q, Cheng X, Zhang Y, Weisman LS, Delling M, Xu H (2010) PI(3,5)P(2) controls membrane trafficking by direct activation of mucolipin Ca(2+) release channels in the endolysosome. Nat Commun 1:38

Dunn T, Gable K, Beeler T (1994) Regulation of cellular Ca2+ by yeast vacuoles. J Biol Chem 269(10): 7273–7278

Durr G, Strayle J, Plemper R, Elbs S, Klee SK, Catty P, Wolf DH, Rudolph HK (1998) The medial-Golgi ion pump Pmr1 supplies the yeast secretory pathway with Ca2+ and Mn2+ required for glycosylation, sorting, and endoplasmic reticulum-associated protein degradation. Mol Biol Cell 9 (5):1149–1162

Ermak G, Davies KJ (2002) Calcium and oxidative stress: from cell signaling to cell death. Mol Immunol 38(10):713–721

Fischer M, Schnell N, Chattaway J, Davies P, Dixon G, Sanders D (1997) The *Saccharomyces cerevisiae* CCH1 gene is involved in calcium influx and mating. FEBS Lett 419(2–3):259–262

Flegg MB, Rudiger S, Erban R (2013) Diffusive spatio-temporal noise in a first-passage time model for intracellular calcium release. J Chem Phys 138 (15):154103

Forster C, Kane PM (2000) Cytosolic Ca2+ homeostasis is a constitutive function of the V-ATPase in *Saccharomyces cerevisiae*. J Biol Chem 275(49): 38245–38253

Ganitkevich VY (2003) The role of mitochondria in cytoplasmic Ca2+ cycling. Exp Physiol 88(1): 91–97

Groenendyk J, Lynch J, Michalak M (2004) Calreticulin, Ca2+, and calcineurin – signaling from the endoplasmic reticulum. Mol Cells 17(3):383–389

Gunter TE, Gunter KK (2001) Uptake of calcium by mitochondria: transport and possible function. IUBMB Life 52(3–5):197–204

Haiech J, Moreau M (2011) The calcium signal: a universal carrier to code, decode and transduce information. Biochimie 93(12):v

Hartwig C, Bahre H, Wolter S, Beckert U, Kaever V, Seifert R (2014) cAMP, cGMP, cCMP and cUMP concentrations across the tree of life: high cCMP and cUMP levels in astrocytes. Neurosci Lett 579:183–187

Hashimoto K, Kudla J (2011) Calcium decoding mechanisms in plants. Biochimie 93(12):2054–2059

He W, Hu Z (2012) The role of the Golgi-resident SPCA Ca(2)(+)/Mn(2)(+) pump in ionic homeostasis and neural function. Neurochem Res 37(3): 455–468

Herchuelz A, Nguidjoe E, Jiang L, Pachera N (2013) Na (+)/Ca(2+) exchange and the plasma membrane Ca(2+)-ATPase in beta-cell function and diabetes. Adv Exp Med Biol 961:385–394

Hernandez-Cruz A, Escobar AL, Jimenez N (1997) Ca(2 +)-induced Ca2+ release phenomena in mammalian sympathetic neurons are critically dependent on the rate of rise of trigger Ca2+. J Gen Physiol 109(2):147–167

Hohendanner F, McCulloch AD, Blatter LA, Michailova AP (2014) Calcium and IP3 dynamics in cardiac myocytes: experimental and computational perspectives and approaches. Front Pharmacol 5:35

Horikawa Y, Goel A, Somlyo AP, Somlyo AV (1998) Mitochondrial calcium in relaxed and tetanized myocardium. Biophys J 74(3):1579–1590

Iida H, Nakamura H, Ono T, Okumura MS, Anraku Y (1994) MID1, a novel *Saccharomyces cerevisiae* gene encoding a plasma membrane protein, is required for Ca2+ influx and mating. Mol Cell Biol 14(12):8259–8271

Ji G, Feldman M, Doran R, Zipfel W, Kotlikoff MI (2006) Ca2+-induced Ca2+ release through localized Ca2+ uncaging in smooth muscle. J Gen Physiol 127(3):225–235

Jiang H, Liu F, Zhang S, Lu L (2014a) Putative PmrA and PmcA are important for normal growth, morphogenesis and cell wall integrity, but not for viability in *Aspergillus nidulans*. Microbiology 160 (Pt 11):2387–2395

Jiang H, Shen Y, Liu W, Lu L (2014b) Deletion of the putative stretch-activated ion channel Mid1 is hypervirulent in *Aspergillus fumigatus*. Fungal Genet Biol 62:62–70

Kakinuma Y, Ohsumi Y, Anraku Y (1981) Properties of H+-translocating adenosine triphosphatase in vacuolar membranes of *Saccharomyces cerevisiae*. J Biol Chem 256(21):10859–10863

Kanzaki M, Nagasawa M, Kojima I, Sato C, Naruse K, Sokabe M, Iida H (1999) Molecular identification of a eukaryotic, stretch-activated nonselective cation channel. Science 285(5429):882–886

Katoh I, Sato S, Fukunishi N, Yoshida H, Imai T, Kurata S (2008) Apaf-1-deficient fog mouse cell apoptosis involves hypo-polarization of the mitochondrial inner membrane, ATP depletion and citrate accumulation. Cell Res 18(12):1210–1219

Kellermayer R, Aiello DP, Miseta A, Bedwell DM (2003) Extracellular Ca(2+) sensing contributes to excess Ca(2+) accumulation and vacuolar fragmentation in a pmr1Delta mutant of *S. cerevisiae*. J Cell Sci 116.(Pt 8:1637–1646

Kmetzsch L, Staats CC, Simon E, Fonseca FL, de Oliveira DL, Sobrino L, Rodrigues J, Leal AL, Nimrichter L, Rodrigues ML, Schrank A, Vainstein MH (2010) The vacuolar Ca(2)(+) exchanger Vcx1 is involved in calcineurin-dependent Ca(2)(+) tolerance and virulence in *Cryptococcus neoformans*. Eukaryot Cell 9(11):1798–1805

Kovacs-Bogdan E, Sancak Y, Kamer KJ, Plovanich M, Jambhekar A, Huber RJ, Myre MA, Blower MD, Mootha VK (2014) Reconstitution of the mitochondrial calcium uniporter in yeast. Proc Natl Acad Sci USA 111(24):8985–8990

Krizanova O, Ondrias K (2003) The inositol 1,4,5-trisphosphate receptor-transcriptional regulation and modulation by phosphorylation. Gen Physiol Biophys 22(3):295–311

Lanner JT, Georgiou DK, Joshi AD, Hamilton SL (2010) Ryanodine receptors: structure, expression, molecular details, and function in calcium release. Cold Spring Harb Perspect Biol 2(11):a003996

Li X, Qian J, Wang C, Zheng K, Ye L, Fu Y, Han N, Bian H, Pan J, Wang J, Zhu M (2011) Regulating cytoplasmic calcium homeostasis can reduce aluminum toxicity in yeast. PLoS One 6(6):e21148

Li LH, Tian XR, Jiang Z, Zeng LW, He WF, Hu ZP (2013) The Golgi apparatus: panel point of cytosolic Ca(2+) regulation. Neurosignals 21(3–4): 272–284

Lissandron V, Podini P, Pizzo P, Pozzan T (2010) Unique characteristics of Ca2+ homeostasis of the trans-Golgi compartment. Proc Natl Acad Sci USA 107(20):9198–9203

Liu M, Du P, Heinrich G, Cox GM, Gelli A (2006) Cch1 mediates calcium entry in *Cryptococcus neoformans* and is essential in low-calcium environments. Eukaryot Cell 5(10):1788–1796

Liu S, Hou Y, Liu W, Lu C, Wang W, Sun S (2015) Components of the calcium-calcineurin signaling pathway in fungal cells and their potential as antifungal targets. Eukaryot Cell 14(4):324–334

Locke EG, Bonilla M, Liang L, Takita Y, Cunningham KW (2000) A homolog of voltage-gated Ca(2+) channels stimulated by depletion of secretory Ca (2+) in yeast. Mol Cell Biol 20(18):6686–6694

Losev E, Papanikou E, Rossanese OW, Glick BS (2008) Cdc1p is an endoplasmic reticulum-localized putative lipid phosphatase that affects Golgi inheritance and actin polarization by activating Ca2+ signaling. Mol Cell Biol 28(10):3336–3343

Loukin S, Zhou X, Kung C, Saimi Y (2008) A genome-wide survey suggests an osmoprotective role for vacuolar Ca2+ release in cell wall-compromised yeast. FASEB J 22(7):2405–2415

Luan S (2009) The CBL-CIPK network in plant calcium signaling. Trends Plant Sci 14(1):37–42

Lustoza AC, Palma LM, Facanha AR, Okorokov LA, Okorokova-Facanha AL (2011) P(5A)-type ATPase Cta4p is essential for Ca2+ transport in the endoplasmic reticulum of *Schizosaccharomyces pombe*. PLoS One 6(11):e27843

Marchi V, Sorin A, Wei Y, Rao R (1999) Induction of vacuolar Ca2+-ATPase and H+/Ca2+ exchange activity in yeast mutants lacking Pmr1, the Golgi Ca2+-ATPase. FEBS Lett 454(3):181–186

Martin DC, Kim H, Mackin NA, Maldonado-Baez L, Evangelista CC Jr, Beaudry VG, Dudgeon DD, Naiman DQ, Erdman SE, Cunningham KW (2011) New regulators of a high affinity Ca2+ influx system revealed through a genome-wide screen in yeast. J Biol Chem 286(12):10744–10754

Matheos DP, Kingsbury TJ, Ahsan US, Cunningham KW (1997) Tcn1p/Crz1p, a calcineurin-dependent transcription factor that differentially regulates gene expression in *Saccharomyces cerevisiae*. Genes Dev 11(24):3445–3458

Micaroni M, Mironov AA (2010) Roles of Ca and secretory pathway Ca-ATPase pump type 1 (SPCA1) in intra-Golgi transport. Commun Integr Biol 3 (6):504–507

Michalak M, Robert Parker JM, Opas M (2002) Ca2+ signaling and calcium binding chaperones of the endoplasmic reticulum. Cell Calcium 32(5–6): 269–278

Michelangeli F, Ogunbayo OA, Wootton LL (2005) A plethora of interacting organellar Ca2+ stores. Curr Opin Cell Biol 17(2):135–140

Miseta A, Fu L, Kellermayer R, Buckley J, Bedwell DM (1999a) The Golgi apparatus plays a significant role in the maintenance of Ca2+ homeostasis in the vps33Delta vacuolar biogenesis mutant of *Saccharomyces cerevisiae*. J Biol Chem 274(9):5939–5947

Miseta A, Kellermayer R, Aiello DP, Fu L, Bedwell DM (1999b) The vacuolar Ca2+/H+ exchanger Vcx1p/Hum1p tightly controls cytosolic Ca2+ levels in *S. cerevisiae*. FEBS Lett 451(2):132–136

Molinaro P, Cantile M, Cuomo O, Secondo A, Pannaccione A, Ambrosino P, Pignataro G, Fiorino F, Severino B, Gatta E, Sisalli MJ, Milanese M, Scorziello A, Bonanno G, Robello M, Santagada V, Caliendo G, Di Renzo G, Annunziato L (2013) Neurounina-1, a novel compound that increases Na+/Ca2+ exchanger activity, effectively protects against stroke damage. Mol Pharmacol 83(1): 142–156

Moreno SN, Docampo R (2009) The role of acidocalcisomes in parasitic protists. J Eukaryot Microbiol 56(3):208–213

Muller EM, Locke EG, Cunningham KW (2001) Differential regulation of two Ca(2+) influx systems by pheromone signaling in *Saccharomyces cerevisiae*. Genetics 159(4):1527–1538

Muller EM, Mackin NA, Erdman SE, Cunningham KW (2003) Fig1p facilitates Ca2+ influx and cell fusion during mating of *Saccharomyces cerevisiae*. J Biol Chem 278(40):38461–38469

Nakano A (2004) Yeast Golgi apparatus–dynamics and sorting. Cell Mol Life Sci 61(2):186–191

Okamura H, Denawa M, Ohniwa R, Takeyasu K (2003) P-type ATPase superfamily: evidence for critical roles for kingdom evolution. Ann N Y Acad Sci 986:219–223

Onyewu C, Wormley FL Jr, Perfect JR, Heitman J (2004) The calcineurin target, Crz1, functions in azole tolerance but is not required for virulence of *Candida albicans*. Infect Immun 72(12):7330–7333

Palmer CP, Zhou XL, Lin J, Loukin SH, Kung C, Saimi Y (2001) A TRP homolog in *Saccharomyces cerevisiae* forms an intracellular Ca(2+)-permeable channel in the yeast vacuolar membrane. Proc Natl Acad Sci USA 98(14):7801–7805

Pan Y, Mansfield KD, Bertozzi CC, Rudenko V, Chan DA, Giaccia AJ, Simon MC (2007) Multiple factors affecting cellular redox status and energy metabolism modulate hypoxia-inducible factor prolyl hydroxylase activity in vivo and in vitro. Mol Cell Biol 27(3):912–925

Parlati F, Dominguez M, Bergeron JJ, Thomas DY (1995) *Saccharomyces cerevisiae* CNE1 encodes an endoplasmic reticulum (ER) membrane protein with sequence similarity to calnexin and calreticulin and functions as a constituent of the ER quality control apparatus. J Biol Chem 270(1):244–253

Perrino BA, Fong YL, Brickey DA, Saitoh Y, Ushio Y, Fukunaga K, Miyamoto E, Soderling TR (1992) Characterization of the phosphatase activity of a baculovirus-expressed calcineurin A isoform. J Biol Chem 267(22):15965–15969

Pfeiffer DR, Gunter TE, Eliseev R, Broekemeier KM, Gunter KK (2001) Release of Ca2+ from mitochondria via the saturable mechanisms and the permeability transition. IUBMB Life 52(3–5): 205–212

Pittman JK (2011) Vacuolar Ca(2+) uptake. Cell Calcium 50(2):139–146

Pivovarova NB, Hongpaisan J, Andrews SB, Friel DD (1999) Depolarization-induced mitochondrial Ca accumulation in sympathetic neurons: spatial and temporal characteristics. J Neurosci 19(15): 6372–6384

Pizzo P, Lissandron V, Capitanio P, Pozzan T (2011) Ca (2+) signalling in the Golgi apparatus. Cell Calcium 50(2):184–192

Prole DL, Taylor CW (2012) Identification and analysis of cation channel homologues in human pathogenic fungi. PLoS One 7(8):e42404

Raeymaekers L, Wuytack E, Willems I, Michiels CW, Wuytack F (2002) Expression of a P-type Ca(2+)-transport ATPase in Bacillus subtilis during sporulation. Cell Calcium 32(2):93

Roome CJ, Empson RM (2013) The contribution of the sodium-calcium exchanger (NCX) and plasma membrane Ca(2+) ATPase (PMCA) to cerebellar synapse function. Adv Exp Med Biol 961:251–263

Rudolph HK, Antebi A, Fink GR, Buckley CM, Dorman TE, LeVitre J, Davidow LS, Mao JI, Moir DT (1989) The yeast secretory pathway is perturbed by mutations in PMR1, a member of a Ca2+ ATPase family. Cell 58(1):133–145

Rueda CB, Llorente-Folch I, Amigo I, Contreras L, Gonzalez-Sanchez P, Martinez-Valero P, Juaristi I, Pardo B, del Arco A, Satrustegui J (2014) Ca (2+) regulation of mitochondrial function in neurons. Biochim Biophys Acta 1837(10):1617–1624

Rusnak F, Mertz P (2000) Calcineurin: form and function. Physiol Rev 80(4):1483–1521

Samanta K, Douglas S, Parekh AB (2014) Mitochondrial calcium uniporter MCU supports cytoplasmic Ca2 + oscillations, store-operated Ca2+ entry and Ca2 +-dependent gene expression in response to receptor stimulation. PLoS One 9(7):e101188

Samarao SS, Teodoro CE, Silva FE, Ribeiro CC, Granato TM, Bernardes NR, Retamal CA, Facanha AR, Okorokova-Facanha AL, Okorokov LA (2009) V H+-ATPase along the yeast secretory pathway: energization of the ER and Golgi membranes. Biochim Biophys Acta 1788(2):303–313

Santo-Domingo J, Demaurex N (2010) Calcium uptake mechanisms of mitochondria. Biochim Biophys Acta 1797(6–7):907–912

Schwaller B (2010) Cytosolic Ca2+ buffers. Cold Spring Harb Perspect Biol 2(11):a004051

Shnyder SD, Mangum JE, Hubbard MJ (2008) Triplex profiling of functionally distinct chaperones (ERp29/PDI/BiP) reveals marked heterogeneity of the endoplasmic reticulum proteome in cancer. J Proteome Res 7(8):3364–3372

Silverman-Gavrila LB, Lew RR (2002) An IP3-activated Ca2+ channel regulates fungal tip growth. J Cell Sci 115.(Pt 24:5013–5025

Song J, Liu X, Zhai P, Huang J, Lu L (2016) A putative mitochondrial calcium uniporter in *A. fumigatus* contributes to mitochondrial Ca(2+) homeostasis and stress responses. Fungal Genet Biol 94:15–22

Squizani ED, Oliveira NK, Reuwsaat JCV, Marques BM, Lopes W, Gerber AL, de Vasconcelos ATR, Lev S, Djordjevic JT, Schrank A, Vainstein MH, Staats CC, Kmetzsch L (2018) Cryptococcal dissemination to the central nervous system requires the vacuolar calcium transporter Pmc1. Cell Microbiol 20(2):e12803

Stemmer PM, Klee CB (1994) Dual calcium ion regulation of calcineurin by calmodulin and calcineurin B. Biochemistry 33(22):6859–6866

Suda Y, Nakano A (2012) The yeast Golgi apparatus. Traffic 13(4):505–510

Takeuchi A, Kim B, Matsuoka S (2015) The destiny of Ca released by mitochondria. J Physiol Sci 65: 11–24

Tanida I, Hasegawa A, Iida H, Ohya Y, Anraku Y (1995) Cooperation of calcineurin and vacuolar H(+)-ATPase in intracellular Ca2+ homeostasis of yeast cells. J Biol Chem 270(17):10113–10119

Ton VK, Mandal D, Vahadji C, Rao R (2002) Functional expression in yeast of the human secretory pathway Ca(2+), Mn(2+)-ATPase defective in Hailey-Hailey disease. J Biol Chem 277(8):6422–6427

Tsai MF, Jiang D, Zhao L, Clapham D, Miller C (2014) Functional reconstitution of the mitochondrial Ca2+/H+ antiporter Letm1. J Gen Physiol 143 (1):67–73

Tuteja N, Mahajan S (2007) Calcium signaling network in plants: an overview. Plant Signal Behav 2(2): 79–85

Ueda Y, Ishitsuka R, Hullin-Matsuda F, Kobayashi T (2014) Regulation of the transbilayer movement of diacylglycerol in the plasma membrane. Biochimie 107:43–50

Uemura S, Kihara A, Iwaki S, Inokuchi J, Igarashi Y (2007) Regulation of the transport and protein levels of the inositol phosphorylceramide mannosyltransferases Csg1 and Csh1 by the Ca2+-binding protein Csg2. J Biol Chem 282(12): 8613–8621

Van Baelen K, Vanoevelen J, Missiaen L, Raeymaekers L, Wuytack F (2001) The Golgi PMR1 P-type ATPase of *Caenorhabditis elegans*. Identification of the gene and demonstration of calcium and manganese transport. J Biol Chem 276(14): 10683–10691

Van Baelen K, Dode L, Vanoevelen J, Callewaert G, De Smedt H, Missiaen L, Parys JB, Raeymaekers L, Wuytack F (2004) The Ca2+/Mn2+ pumps in the Golgi apparatus. Biochim Biophys Acta 1742 (1–3):103–112

von Blume J, Alleaume AM, Cantero-Recasens G, Curwin A, Carreras-Sureda A, Zimmermann T, van Galen J, Wakana Y, Valverde MA, Malhotra V (2011) ADF/cofilin regulates secretory cargo sorting at the TGN via the Ca2+ ATPase SPCA1. Dev Cell 20(5):652–662

Wada Y, Anraku Y (1994) Chemiosmotic coupling of ion transport in the yeast vacuole: its role in acidification inside organelles. J Bioenerg Biomembr 26 (6):631–637

Wang S, Cao J, Liu X, Hu H, Shi J, Zhang S, Keller NP, Lu L (2012) Putative calcium channels CchA and MidA play the important roles in conidiation, hyphal polarity and cell wall components in *Aspergillus nidulans*. PLoS One 7(10):e46564

White PJ, Broadley MR (2003) Calcium in plants. Ann Bot 92(4):487–511

Wuytack F, Raeymaekers L, Missiaen L (2003) PMR1/SPCA Ca2+ pumps and the role of the Golgi apparatus as a Ca2+ store. Pflugers Arch 446 (2):148–153

Yagodin S, Pivovarova NB, Andrews SB, Sattelle DB (1999) Functional characterization of thapsigargin and agonist-insensitive acidic Ca2+ stores in *Drosophila melanogaster* S2 cell lines. Cell Calcium 25 (6):429–438

Yang M, Brand A, Srikantha T, Daniels KJ, Soll DR, Gow NA (2011) Fig1 facilitates calcium influx and localizes to membranes destined to undergo fusion during mating in *Candida albicans*. Eukaryot Cell 10(3):435–444

Yu Q, Wang F, Zhao Q, Chen J, Zhang B, Ding X, Wang H, Yang B, Lu G, Li M (2014) A novel role of the vacuolar calcium channel Yvc1 in stress response, morphogenesis and pathogenicity of *Candida albicans*. Int J Med Microbiol 304(3–4):339–350

Zhang S, Zheng H, Long N, Carbo N, Chen P, Aguilar PS, Lu L (2014) FigA, a putative homolog of low-affinity calcium system member Fig1 in *Saccharomyces cerevisiae*, is involved in growth and asexual and sexual development in *Aspergillus nidulans*. Eukaryot Cell 13(2):295–303

Zhang Y, Zheng Q, Sun C, Song J, Gao L, Zhang S, Munoz A, Read ND, Lu L (2016) Palmitoylation of the cysteine residue in the DHHC motif of a palmitoyl transferase mediates Ca2+ homeostasis in *Aspergillus*. PLoS Genet 12(4):e1005977

Zhou XL, Batiza AF, Loukin SH, Palmer CP, Kung C, Saimi Y (2003) The transient receptor potential channel on the yeast vacuole is mechanosensitive. Proc Natl Acad Sci USA 100(12):7105–7110

Biosystematic Index

Biology of the Fungal Cell, 3rd Edition
The Mycota VIII
D. Hoffmeister, M. Gressler (Eds.)

Subject Index

Biology of the Fungal Cell, 3rd Edition
The Mycota VIII
D. Hoffmeister, M. Gressler (Eds.)

Printed by Printforce, the Netherlands